IEE CONTROL ENGINEERING SERIES 61

Series Editors: Professor D. P. Atherton
Professor G. W. Irwin

Nonlinear predictive control

theory and practice

Other volumes in print in this series:

Volume 2 **Elevator traffic analysis, design and control** G. C. Barney and S. M. dos Santos
Volume 8 **A history of control engineering, 1800-1930** S. Bennett
Volume 14 **Optimal relay and saturating control system synthesis** E. P. Ryan
Volume 15 **Self-tuning and adaptive control: theory and application** C. J. Harris and S. A. Billings (Editors)
Volume 16 **Systems modelling and optimisation** P. Nash
Volume 18 **Applied control theory** J. R. Leigh
Volume 19 **Stepping motors: a guide to modern theory and practice** P. P. Acarnley
Volume 20 **Design of modern control systems** D. J. Bell, P. A. Cook and N. Munro (Editors)
Volume 21 **Computer control of industrial processes** S. Bennett and D. A. Linkens (Editors)
Volume 23 **Robotic technology** A. Pugh (Editor)
Volume 26 **Measurement and instrumentation for control** M. G. Mylroi and G. Calvert (Editors)
Volume 27 **Process dynamics estimation and control** A. Johnson
Volume 28 **Robots and automated manufacture** J. Billingsley (Editor)
Volume 30 **Electromagnetic suspension—dynamics and control** P. K. Slnha
Volume 31 **Modelling and control of fermentation processes** J. R. Leigh (Editor)
Volume 32 **Multivariable control for industrial applications** J. O'Reilly (Editor)
Volume 34 **Singular perturbation methodology in control systems** D. S. Naidu
Volume 35 **Implementation of self-tuning controllers** K. Warwick (Editor)
Volume 36 **Robot control** K. Warwick and A. Pugh (Editors)
Volume 37 **Industrial digital control systems (revised edition)** K. Warwick and D. Rees (Editors)
Volume 38 **Parallel processing in control** P. J. Fleming (Editor)
Volume 39 **Continuous time controller design** R. Balasubramanian
Volume 40 **Deterministic control of uncertain systems** A. S. I. Zinober (Editor)
Volume 41 **Computer control of real-time processes** S. Bennett and G. S. Virk (Editors)
Volume 42 **Digital signal processing: principles, devices and applications** N. B. Jones and J. D. McK. Watson (Editors)
Volume 43 **Trends in information technology** D. A. Linkens and R. I. Nicolson (Editors)
Volume 44 **Knowledge-based systems for industrial control** J. McGhee, M. J. Grimble and A. Mowforth (Editors)
Volume 45 **Control theory—a guided tour** J. R. Leigh
Volume 46 **Neural networks for control and systems** K. Warwick, G. W. Irwin and K. J. Hunt (Editors)
Volume 47 **A history of control engineering, 1930-1956** S. Bennett
Volume 48 **MATLAB toolboxes and applications for control** A. J. Chipperfield and P. J. Fleming (Editors)
Volume 49 **Polynomial methods in optimal control and filtering** K. J. Hunt (Editor)
Volume 50 **Programming industrial control systems using IEC 1131-3** R. W. Lewis
Volume 51 **Advanced robotics and intelligent machines** J. O. Gray and D. G. Caldwell (Editors)
Volume 52 **Adaptive prediction and predictive control** P. P. Kanjilal
Volume 53 **Neural network applications in control** G. W. Irwin, K. Warwick and K. J. Hunt (Editors)
Volume 54 **Control engineering solutions: a practical approach** P. Albertos, R. Strietzel and N. Mort (Editors)
Volume 55 **Genetic algorithms in engineering systems** A. M. S. Zalzala and P. J. Fleming (Editors)
Volume 56 **Symbolic methods in control system analysis and design** N. Munro (Editor)
Volume 57 **Flight control systems** R. W. Pratt (Editor)
Volume 58 **Power-plant control and instrumentation** D. Lindsley
Volume 59 **Modelling control systems using IEC 61499** R. Lewis
Volume 60 **People in control: human factors in control room design** J. Noyes and M. Bransby (Editors)

Nonlinear predictive control

theory and practice

Edited by

Basil Kouvaritakis and
Mark Cannon

The Institution of Electrical Engineers

Published by: The Institution of Electrical Engineers, London,
United Kingdom

The Institution of Electrical Engineers,
Michael Faraday House,
Six Hills Way, Stevenage,
Herts. SG1 2AY, United Kingdom

British Library Cataloguing in Publication Data

Advances in non-linear model predictive control.—
(IEE control engineering series; no.61)
1. Predictive control 2. Nonlinear control theory
I. Kouvaritakis, Basil II. Cannon, Mark III. Institution of Electrical Engineers
629.8

ISBN 0 85296 984 8

Typeset by Tradespools Ltd, Frome, Somerset
Printed in England by Antony Rowe Ltd, Chippenham, Wiltshire

Contents

Preface

Model predictive control has, for several decades, been a fertile area of research but above all has proved enormously successful in industry, especially in the context of process control. The key to its popularity is its ability to take systematic account of constraints, thereby allowing processes to operate at the limits of achievable performance. In terms of linear models the field has reached maturity, as evidenced by the appearance in the literature of a plethora of survey papers and books.

Given that the dynamics of most real plant are nonlinear, it was natural for researchers to ask whether the benefits of linear MPC could be transferred to the nonlinear case. This has presented some challenging problems, both theoretical and practical, but the last decade has seen the emergence of some significant results, especially in respect of guaranteeing closed-loop stability. There now exists a whole range of techniques, some of which make use of dual mode predictions in conjunction with terminal penalties and/or control Lyapunov functions, positively invariant sets and terminal stability constraints, others which deploy feedback linearisation or differential flatness. The reader will find examples of all these in Parts I and II of this book. However, theory alone cannot establish nonlinear model predictive control (NMPC) as an industrial standard and the book goes on to look at some of the other key issues, such as computation, optimality and modelling.

Online computational complexity is a major concern in NMPC, especially for fast sampling applications, high dimensional systems and control problems that demand the use of large prediction horizons. The performance costs and constraints are in general nonconvex functions of the predicted inputs, and their optimisation calls for the use of numerical techniques, whose demanding nature may exceed the time available for online computation. It therefore becomes essential to look for suboptimal solutions. Parts II and III discuss a range of suboptimal approaches based on linearisation about predicted trajectories, feedback linearisation, interpolation, and approximations to optimal control Lyapunov functions.

Inextricably connected with approximation is the question of the degree of suboptimality, and a quantitative analysis of this is presented in some of the chapters of Part II, both in terms of the open- and closed-loop costs. Various approaches are discussed, some deploying upper bounds on the cost, others invoking inverse optimality, or measuring the distance between optimal and approximate value functions.

The success of MPC depends to a large extent on the availability of reliable models, and for the case of nonlinear plant this can be quite challenging. On the one hand phenomenological modelling can be expensive and may lead to unnecessarily complicated system descriptions; on the other, empirical input-output descriptions require appropriate selection of model structures, test signals and validation procedures. These questions are considered in detail in Parts I and IV, which propose systematic means of modelling classes of processes via neural networks. The efficacy of such models and the related MPC strategies are demonstrated in terms of applications such as IC engine control, pH neutralisation control, acid recovery process control, distillation dual composition control and control of a nonisothermal CSTR.

The four parts of the book are meant to be distinctive but inevitably overlap with each other to a limited extent. The first part comprises two chapters of wide scope that survey a number of theoretical and practical trends within the field. The material of the second part will appeal mostly to the theoretician, although in each chapter the theory is demonstrated in the form of a practical application. The main concern of the third part is the derivation of NMPC strategies which provide the appropriate guarantees for closed-loop stability, but in addition trade off a certain degree of optimality in return for a significant gain in computational efficiency. The emphasis of the final part is mainly on the practical implementation of NMPC.

By tackling a range of contentious issues, the book attempts to bring together all of the components whose synergy is a prerequisite for the future success of NMPC. Theoretical rigour, reliable modelling, computational efficiency, *a priori* assessment of benefits in terms of performance, and a broad sample of successful applications are all needed to convert the sceptic, and to justify the cost of switching from MPC to NMPC. It is not, of course, claimed that NMPC will eclipse linear MPC altogether, but rather that NMPC has tremendous potential for a wide range of industrial problems. It is hoped that the material of this book provides irrefutable evidence of this and points the way forward.

Basil Kouvaritakis
Mark Cannon

Oxford University
Department of Engineering Science

Contributors

Chapter 1

T.A. Badgwell
Advanced Technology Group, Aspen Technology, Inc., 1293 Eldridge Parkway, Houston, TX 77077, USA

S.J. Qin
Department of Chemical Engineering, The University of Texas at Austin, Austin, TX 78712, USA

Chapter 2

R.S. Parker, E.P. Gatzke, R. Mahadevan, E.S. Meadows and F.J. Doyle III
Department of Chemical Engineering, University of Delaware, Newark, DE 19716, USA

Chapter 3

L. Magni, G. De Nicolao and R. Scattolini
Dipartimento di Informatica e Sistemistica, University of Pavia, Via Ferrata 1, 27100 Pavia, Italy

Chapter 4

M. Sznaier
Department of Electrical Engineering, Penn State University, University Park, PA 16802, USA

J. Cloutier
Navigation and Control Branch, Air Force Research Laboratory, Eglin AFB, FL 32542, USA

Chapter 5

M. Niemiec
Honeywell Inc., 16404 N. Black Canyon Hwy, Phoenix, AZ 85023, USA

C. Kravaris
Department of Chemical Engineering, University of Patras, GR-26500, Patras, Greece

Chapter 6

M. Cannon and B. Kouvaritakis
Department of Engineering Science, University of Oxford, Parks Road, Oxford, OX1 3PJ, UK

Chapter 7

B. Kouvaritakis, J.A. Rossiter and M. Cannon
Department of Engineering Science, University of Oxford, Parks Road, Oxford, OX1 3PJ, UK

Chapter 8

A. Zheng
Department of Chemical Engineering, University of Massachusetts, Amherst, MA 01003, USA

Chapter 9

M. Soroush
Department of Chemical Engineering, Drexel University, Philadelphia, PA 19104, USA

Chapter 10

B.A. Ogunnaike
E. I. Du Pont de Nemours and Company, Experimental Station E1/102, Wilmington, DE 19880, USA

Chapter 11

S. Townsend
Anex6 Ltd, NISoft House, Ravenhill Road, Belfast BT6 8AW, UK

G.W. Irwin
Intelligent Systems and Control Group, Department of Electrical and Electronic Engineering, The Queen's University of Belfast, Belfast BT9 5AH, UK

Chapter 12

B. Lennox
School of Engineering, Simon Building, University of Manchester, Oxford Road, Manchester M13 9PL, UK

G. Montague
Department of Chemical and Process Engineering, Merz Court, University of Newcastle, Newcastle-upon-Tyne NE1 7RU, UK

Part I

Chapter 1

Review of nonlinear model predictive control applications

Thomas A. Badgwell and S. Joe Qin

Abstract

This chapter provides an overview of nonlinear model predictive control (NMPC) applications in industry, focusing primarily on recent applications reported by NMPC vendors. A brief summary of NMPC theory is presented to highlight issues pertinent to NMPC applications. Several industrial NMPC implementations are then discussed with reference to modelling, control, optimisation and implementation issues. Results from several industrial applications are presented to illustrate the benefits possible with NMPC technology. The chapter concludes with a discussion of future needs in NMPC theory and practice.

1.1 Introduction

The term 'model predictive control' (MPC) refers to a class of computer control algorithms that utilise an explicit process model to predict the future response of a plant. At each control interval an MPC algorithm determines a sequence of manipulated variable adjustments that optimise future plant behaviour. The first input in the optimal sequence is then sent into the plant, and the entire optimisation is repeated at subsequent control intervals. MPC technology was originally developed for power plant and petroleum refinery control applications, but can now be found in a wide variety of manufacturing environments including chemicals, food processing, automotive, aerospace, metallurgy and pulp and paper.

Theoretical and practical issues associated with MPC technology are summarised in several recent publications. Rawlings [1] provides an excellent introduction to

MPC technology aimed at the nonspecialist. Qin and Badgwell [2] present a history of MPC technology development and a survey of industrial applications, focused primarily on those employing linear models. Allgöwer *et al.* [3] provide a high-level introduction to moving horizon estimation and model predictive control using nonlinear models. Mayne *et al.* [4] summarise the most recent theoretical efforts to understand closed-loop properties of MPC algorithms.

The success of MPC technology as a process control paradigm can be attributed to three important factors. First and foremost is the incorporation of an explicit process model into the control calculation. This allows the controller, in principle, to deal directly with all significant features of the process dynamics. Second, the MPC algorithm considers plant behaviour over a future horizon in time. This means that the effects of feedforward and feedback disturbances can be anticipated and removed, allowing the controller to drive the plant more closely along a desired future trajectory. Finally, the MPC controller considers process input, state and output constraints directly in the control calculation. This means that constraint violations are far less likely, resulting in tighter control at the optimal constrained steady-state for the process. It is the inclusion of constraints that most clearly distinguishes MPC from other process control paradigms.

Although manufacturing processes are inherently nonlinear, the vast majority of MPC applications to date are based on linear dynamic models, the most common being step and impulse response models derived from the convolution integral. There are several potential reasons for this. Linear empirical models can be identified in a straightforward manner from process test data. In addition, most applications to date have been in refinery processing [2], where the goal is largely to maintain the process at a desired steady state (regulator problem), rather than moving rapidly from one operating point to another (servo problem). A carefully identified linear model is sufficiently accurate for such applications, especially if high-quality feedback measurements are available. Finally, by using a linear model and a quadratic objective, the nominal MPC algorithm takes the form of a highly structured convex quadratic program (QP), for which reliable solution algorithms and software can easily be found [5]. This is important because the solution algorithm must converge reliably to the optimum in no more than a few tens of seconds to be useful in manufacturing applications. For these reasons, in many cases a linear model will provide the majority of the benefits possible with MPC technology.

Nevertheless, there are cases where nonlinear effects are significant enough to justify the use of nonlinear model predictive control (NMPC) technology, which we define here as MPC using a nonlinear model. These include at least two broad categories of applications:

1. regulator control problems where the process is highly nonlinear and subject to large, frequent disturbances (pH control, etc.)
2. servo control problems where the operating points change frequently and span a sufficiently wide range of nonlinear process dynamics (polymer manufacturing, ammonia synthesis, etc.).

It is interesting to note that some of the very first MPC papers describe ways to address nonlinear process behaviour while still retaining a linear dynamic model in the control algorithm. Richalet *et al.* [6], for example, describe how nonlinear behaviour due to load changes in a steam power plant application was handled by executing their identification and command (IDCOM) algorithm at a variable frequency. Prett and Gillette [7] describe applying a dynamic matrix control (DMC) algorithm to control a fluid catalytic cracking unit. Model gains were obtained at each control iteration by perturbing a detailed nonlinear steady-state model. The updated gains were imposed on constant linear dynamics for use in the control calculation.

In a previous survey of MPC technology [2], over 2200 commercial applications were reported. However, almost all of these were implemented with linear models and were clustered in refinery and petrochemical processes. In preparing a more recent survey [8] the authors found a sizeable number of NMPC applications in areas where MPC has not traditionally been applied. Figure 1.1 shows a rough distribution of the number of MPC applications versus the degree of process nonlinearity. MPC technology has not yet penetrated deeply into areas where process nonlinearities are strong and market demands require frequent changes in operating conditions. It is these areas that provide the greatest opportunity for NMPC applications.

While theoretical aspects of NMPC algorithms have been discussed quite effectively in several recent publications (see, for example, Reference 4), descriptions of industrial NMPC applications are more difficult to find. The primary purpose of this chapter is to provide a snapshot of the current state of the

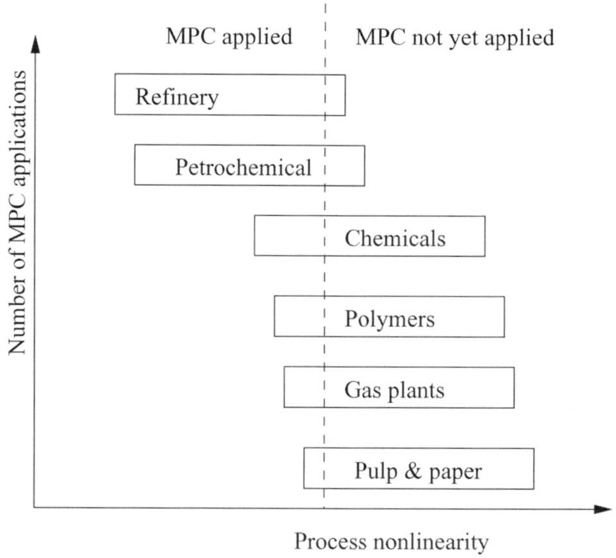

Figure 1.1 Distribution of MPC applications versus degree of process nonlinearity

art in NMPC applications. A brief summary of NMPC theory is presented to highlight what is known about closed-loop properties and to emphasise issues pertinent to NMPC applications. Then several industrial NMPC products are discussed with reference to modelling, control, optimisation and implementation issues. The focus here is on NMPC products that are either commercially available at the present time or were available in the recent past, since these are the implementations that have had the widest impact on NMPC practice. A few illustrative industrial applications are then discussed in detail. The chapter concludes with a discussion of future needs and trends in NMPC theory and applications.

1.2 Theoretical foundations of NMPC

To establish a framework for comparing various NMPC formulations, we first define a simplified NMPC algorithm and then briefly summarise its theoretical properties. The calculations necessary for an implementable MPC algorithm are described in greater detail in Reference 2. For a more complete discussion of theoretical issues pertaining to NMPC the reader is referred to recent review articles by Mayne *et al.* [4] and Allgöwer *et al.* [3].

Assume that the plant to be controlled can be described by the following discrete-time, nonlinear, state-space model:

$$\mathbf{x}_{k+1} = \mathbf{f}(\mathbf{x}_k, \mathbf{u}_k, \mathbf{v}_k, \mathbf{w}_k) \tag{1a}$$

$$\mathbf{y}_k = \mathbf{g}(\mathbf{x}_k) + \xi_k \tag{1b}$$

where $\mathbf{u}_k \in \Re^{m_u}$ is a vector of m_u process inputs or manipulated variables (MVs), $\mathbf{y}_k \in \Re^{m_y}$ is a vector of m_y process outputs or controlled variables (CVs), $\mathbf{x}_k \in \Re^n$ is a vector of n state variables, $\mathbf{v}_k \in \Re^{m_v}$ is a vector of m_v measured disturbance variables (DVs), $\mathbf{w}_k \in \Re^{m_w}$ is a vector of m_w unmeasured DVs or noise and $\xi_k \in \Re^{m_y}$ is a vector of measurement noise.

The control problem to be solved is to compute a sequence of inputs $\{\mathbf{u}_{k+j}\}$ that will take the process from its current state $\boldsymbol{x}_k$ to a desired steady state $\boldsymbol{x}_s$. The desired steady state $(\boldsymbol{y}_s, \boldsymbol{x}_s, \boldsymbol{u}_s)$ is determined by a local steady-state optimisation, which may be based on an economic objective. The optimal steady state must be recalculated at each time step because disturbances entering the plant may change the location of the optimal operating point. Feedforward disturbances are removed by incorporating their effects into the model $\boldsymbol{f}$. Feedback disturbances are typically handled by assuming that a step disturbance has entered at the output and that it will remain constant for all future time. To accomplish this, a bias term that compares the current predicted output $\boldsymbol{y}_k$ to the current measured output $\mathbf{y}_k^m$ is computed:

$$\mathbf{b}_k = \mathbf{y}_k^m - \mathbf{y}_k \tag{2}$$

The bias $\boldsymbol{b}_k$ term is added to the model for use in subsequent predictions:

$$\mathbf{y}_{k+j} = \mathbf{g}(\mathbf{x}_{k+j}) + \mathbf{b}_k \tag{3}$$

The NMPC control algorithms described in this chapter solve a nonlinear program of the form

$$\min J(\mathbf{u}^M) = \sum_{j=1}^{P} \left\|\mathbf{e}^y_{k+j}\right\|^q_{\mathbf{Q}_j} + \left\|\mathbf{s}_j\right\|^q_{\mathbf{T}_j} + \sum_{j=0}^{M-1} \left\|\mathbf{e}^q_{k+j}\right\|^q_{\mathbf{R}_j} + \left\|\Delta\mathbf{u}_{k+j}\right\|^q_{\mathbf{S}_j} \tag{4a}$$

subject to a model constraint:

$$\begin{aligned} \mathbf{x}_{k+j} &= \mathbf{f}(\mathbf{x}_{k+j-1}, \mathbf{u}_{k+j-1}) & \forall j = 1, P \\ \mathbf{y}_{k+j} &= \mathbf{g}(\mathbf{x}_{k+j}) + \mathbf{b}_k & \forall j = 1, P \end{aligned} \tag{4b}$$

and subject to inequality constraints:

$$\begin{aligned} \underline{\mathbf{y}}_j - \mathbf{s}_j \leq \mathbf{y}_{k+j} &\leq \overline{\mathbf{y}}_j + \mathbf{s}_j & \forall j = 1, P \\ \underline{\mathbf{u}} \leq \mathbf{u}_{k+j} &\leq \overline{\mathbf{u}} & \forall j = 0, M-1 \\ \Delta\underline{\mathbf{u}} \leq \Delta\mathbf{u}_{k+j} &\leq \Delta\overline{\mathbf{u}} & \forall j = 0, M-1 \\ \mathbf{s} &\geq 0 \end{aligned} \tag{4c}$$

The objective function in (4a) includes four conflicting contributions. Future output behaviour is controlled by penalising deviations from the desired steady-state $\boldsymbol{y}_s$, defined as $\mathbf{e}^y_{k+j} = \mathbf{y}_{k+j} - \mathbf{y}_s$, over a prediction horizon of length P. Output constraint violations are penalised by minimising the size of output constraint slack variables $\boldsymbol{s}_j$. Future input deviations from the desired steady-state input $\boldsymbol{u}_s$ are controlled using input penalties defined as $\mathbf{e}^u_{k+j} = \mathbf{u}_{k+j} - \mathbf{u}_s$, over a control horizon of length M. Rapid input changes are penalised with a separate term involving the moves $\Delta\mathbf{u}_{k+j}$. The magnitudes of the tracking deviations and constraint violations are measured by vector norms, usually either an L_1 or L_2 norm $(q = 1, 2)$. The relative importance of the objective function contributions is controlled by setting the time dependent weight matrices $\mathbf{Q_j}$, $\mathbf{T_j}$, $\mathbf{S_j}$ and $\mathbf{R_j}$; these are chosen to be positive definite. The solution is a set of M input adjustments:

$$\mathbf{u}^M = (\mathbf{u}_k \quad \mathbf{u}_{k+1} \quad \ldots \quad \mathbf{u}_{k+M-1}) \tag{5}$$

The first input $\boldsymbol{u}_k$ is injected into the plant and the calculation is repeated at the next sample time. The fact that one can solve the closed-loop control problem through a sequence of open-loop optimisations was recognised very early in the development of optimal control theory [9]. One can view the NMPC solution as a way of turning an intractable closed-loop computation into a sequence of tractable open-loop calculations [4].

In principle the NMPC method is limited to those problems for which a globally optimal solution can be found for the nonlinear program (4). The time available for the calculation is generally a small fraction of the control execution interval. With a linear model and a quadratic objective, the resulting optimisation problem takes the form of a highly structured convex quadratic program (QP) for which there exists a unique optimal solution. Several reliable standard solution codes are available for this problem. Introduction of a nonlinear model leads, in the general case, to loss of convexity. This means that it is much more difficult to find a solution and once found, it cannot be guaranteed to be globally optimal. Scokaert *et al.* have shown, however, that for a properly formulated NMPC algorithm, nominal stability (closed-loop stability when the model is perfect and there are no disturbances) can be retained even if a global solution is not available [10].

It was also recognised early in the development of optimal control theory that, no matter how the control problem is solved, optimality does not necessarily imply closed-loop stability, even when the model represents the true plant perfectly [11]. Under certain conditions, however, this problem can be overcome through proper construction of the NMPC algorithm. The decoupling of optimality and closed-loop stability is an issue that is still not widely appreciated by industrial practitioners.

In theory, the most straightforward way to modify NMPC algorithm (4) to achieve nominal stability involves setting the prediction and control horizons to infinity $(P, M \to \infty)$ [3]. With standard technical assumptions, it follows directly from Bellman's Principle of Optimality [12] that the predicted open-loop input and state trajectories will match those achieved in the closed-loop. This implies nominal stability because any feasible trajectory terminates at the desired steady-state.

From a practical point of view, however, it is simply not possible to solve the NMPC optimisation with infinite horizons for a realistic problem. The focus of recent research efforts has been to obtain a computationally tractable approximation of the infinite horizon problem that still retains desirable closed-loop properties. An early solution proposed by Keerthi and Gilbert [13] involves adding a terminal state constraint to the NMPC algorithm of the form

$$\mathbf{x}_{k+P} = \mathbf{x}_s \tag{6}$$

With such a constraint enforced, the objective function for the controller (4a) becomes a Lyapunov function for the closed-loop system, leading to nominal stability. Unfortunately such a constraint may be quite difficult to satisfy in real time; exact satisfaction requires an infinite number of iterations for the numerical solution code. This motivated Michalska and Mayne [14] to seek a less stringent stability requirement. Their main idea is to define a neighbourhood W around the desired steady-state $\boldsymbol{x}_s$ within which the system can be steered to $\boldsymbol{x}_s$ by a constant linear feedback controller. They add to the NMPC algorithm a constraint of the form

$$(\mathbf{x}_{k+P} - \mathbf{x}_s) \in W \tag{7}$$

If the current state $\boldsymbol{x}_k$ lies outside this region then the NMPC algorithm described above is solved with constraint 7. Once inside the region W the control switches to the previously determined constant linear feedback controller. Michalska and Mayne describe this as a dual-mode controller.

Most recent research activity is focused on quasi-infinite horizon NMPC algorithms, first introduced by Chen and Allgöwer [15]. The basic idea motivating this method is similar to that of dual-mode control. The terminal constraint 7 is imposed so that, at the end of the finite horizon $j = P$, one can imagine that a linear stabilising controller takes over. An upper bound for the objective function from $j = P + 1, \infty$ can then be computed, and this term is added as a terminal penalty to the original finite horizon objective. This modified objective is then used regardless of where the current state lies, so that it is not necessary to switch from one controller to another.

These theoretical results provide a foundation upon which to build an implementable NMPC controller. The challenge of the industrial practitioner is to take these ideas to the market place, which means that a number of additional practical issues must be confronted. Among other things, one must choose an appropriate model form, decide how best to identify or derive the model, and develop a reliable numerical solution method. The following section describes how five NMPC vendors have addressed these issues.

1.3 Industrial implementations of NMPC

In this section we describe the control algorithms used in several commercial NMPC products. Table 1.1 lists the products that we examined and the companies that supplied them. Although this list is by no means exhaustive, we believe that the technology sold by these companies is representative of the current state of the art.

Table 1.2 provides information on the details of each algorithm, including the model types used, options at each step in the control calculation, and the optimisation algorithm used to compute the solution. The control algorithm entries correspond to the steps of the simplified NMPC control calculation illustrated in Figure 1.2. The following subsections describe these aspects in greater detail.

1.3.1 Models

The first issue encountered in NMPC implementation is the derivation of a dynamic nonlinear model suitable for model predictive control. In the general practice of

Table 1.1 NMPC companies and product names

Company	*Product name (acronym)*
Adersa	Predictive Functional Control (PFC)
Aspen Technology	Aspen Target
Continental Controls	Multivariable Control (MVC)

Table 1.2 Comparison of industrial NMPC control technology

Company	*Adersa*	*Aspen Technology*	*Continental Controls*	*DOT Products*	*Pavilion Technologies*
Product	PFC	Aspen Target	MVC	NOVA NLC	Process Perfecter
Model forms[a]	NSS-FP	NSS	SNP-IO	NSS-FP	NNN-IO
	S,I,U	S,I,U	S	S,I	S,I,U
Feedback[b]	CD,ID	CD,ID,EKF	CD	CD	CD,ID
SS Opt Obj[c]	Q[I,O]	Q[I,O]	Q[I,O]	–	Q[I,O]
SS Opt Const[d]	IH,OH	IH,OH	IH,OS	–	IH,OH,OS
Dyn Opt Obj[e]	Q[I,O]	Q[I,O,M]	Q[I,O,M]	(Q,A)[I,O,M]	Q[I,O]
Dyn Opt Const[f]	IC,OH	IH,OS-l1	IH,OS	IH,OH,OS	IH,OS
Output Traj[g]	S,Z,RT	S,Z,RT	S,Z,RT	S,Z,RT	S,Z,TW
Output Horiz[h]	FH,CP	FH,CP	FH	FH	FH
Input Param[i]	BF,SM	MM	SM	MM	MM
Sol. method[j]	NLS	MSN-(QPKWIK)	GRG (GRG2)	MCNLP (Nova)	GRG (GRG2)
References		16, 17, 25 26, 31, Note 2	32, 37–39	28, 33	18–20, 29, 40

[a] Model form: (IO) input-output, (FP) first-principles, (NSS) nonlinear state-space, (NNN) nonlinear neural net, (SNP) static nonlinear polynomial, (S) stable, (I) integrating, (U) unstable.
[b] Feedback: (CD) constant output disturbance, (ID) integrating output disturbance, (EKF) extended Kalman filter.
[c] Steady-state optimisation objective: (Q) quadratic, (I) inputs, (O) outputs.
[d] Steady-state optimisation constraints: (IH) input hard maximum, minimum and rate of change constraints, (OH) output hard maximum and minimum constraints.
[e] Dynamic optimisation objective: (Q) quadratic, (A) one norm, (I) inputs, (O) outputs, (M) input moves.
[f] Dynamic optimisation constraints: (IH) input hard maximum, minimum and rate of change constraints, (IC) input clipped maximum, minimum and rate of change constraints, (OH) output hard maximum and minimum constraints, (OS) output soft maximum and minimum constraints, (OS-l1) output soft constraints with l_1 exact penalty treatment [25].
[g] Output trajectory: (S) setpoint, (Z) zone, (RT) reference trajectory, (TW) trajectory weighting.
[h] Output horizon: (FH) finite horizon, (CP) coincidence points.
[i] Input parameterisation: (SM) single move, (MM) multiple move, (BF) basis functions.

linear MPC, the majority of dynamic models are derived from plant testing and system identification. For NMPC, however, the issue of plant testing and system identification becomes much more complicated. In this subsection we present process modelling methods used in the industrial practice of NMPC, which include system identification methods and first principles approaches.

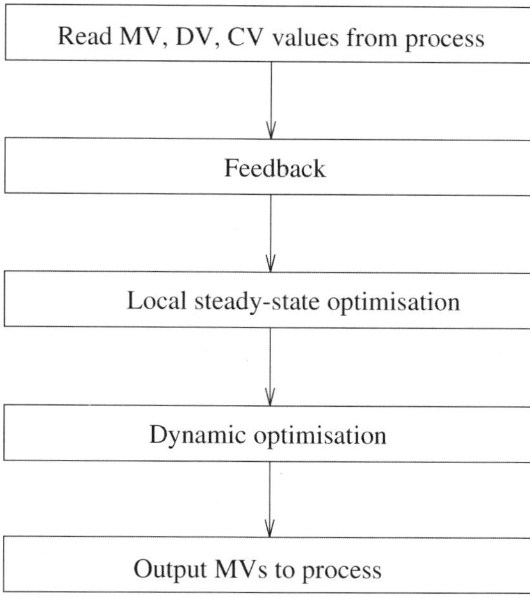

Figure 1.2 A general NMPC control calculation: MV, manipulated variable; DV, disturbance variable; CV, controlled variable

1.3.1.1 State-space models

Because step response and impulse response models are nonparsimonic, a class of state-space models is adopted in the Aspen Target[1] product, which has a linear dynamic state equation and a nonlinear output relation:

$$\mathbf{x}_{k+1} = \mathbf{A}\mathbf{x}_k + \mathbf{B}_u\mathbf{u}_k + \mathbf{B}_v\mathbf{v}_k \tag{8}$$

$$\mathbf{y}_k = \mathbf{g}(\mathbf{x}_k) \tag{9}$$

Here the MVs and DVs can have predetermined time delays. More specifically, the output nonlinearity is modelled as a linear relation superimposed with a nonlinear neural network, that is,

$$\mathbf{g}(\mathbf{x}_k) = \mathbf{C}\mathbf{x}_k + \mathrm{N}(\mathbf{x}_k) \tag{10}$$

Since the state vector $\mathbf{x}$ is not necessarily limited to physical variables, this nonlinear model appears to be more general than measurement nonlinearity. For example, a Wiener model with a dynamic linear model followed by a static nonlinear mapping can be represented in this form. It is claimed that this type of nonlinear model can approximate any discrete time nonlinear processes with fading memory [16].

In nonlinear modelling, selection of a robust and reliable identification algorithm is a more difficult issue than selecting the underlying nonlinear relation. The identification algorithm discussed in Zhao *et al.* [17] builds one model for each output separately. For a process having m_y output variables, overall m_y MISO submodels are built. The following procedure is employed to identify each submodel from process data:

1. Specify a rough time constant for each input-output pair; then a series of first order filters or a Laguerre model is constructed for each input [16, 17]. The filter states for all inputs comprise the state vector **x**.
2. A static linear model is built for each output $\{y_j, j = 1, 2, \ldots, My\}$ using the state vector **x** as inputs using partial least squares (PLS).
3. Model reduction is then performed on the input-state-output model identified in steps 1 and 2 using principal component analysis and internal balancing to eliminate highly collinear state variables.
4. The reduced model is rearranged in a state-space model $(\mathbf{A}, \mathbf{B})$, which is used to generate the state sequence $\{\mathbf{x}_k,\ k = 1, 2, \ldots, K\}$. If the model converges, i.e. there is no further reduction in model order, go to the next step; otherwise, return to step 2.
5. A PLS model is built between the state vector **x** and the output y_j. The PLS model coefficients form the **C** matrix.
6. A neural network model is built between the PLS latent factors in the previous step and the PLS residual of the output y_j. This step generates the nonlinear static map $g_j(\mathbf{x})$. PLS latent factors are used instead of the state vectors to improve the robustness of the neural network training and reduce the size of the neural network.

In a recently submitted article, Turner and Guiver[2] of Aspen Technology describe the potential pitfalls of using neural networks for nonlinear control. The main problem is that for a typical neural network, the model derivatives fall to zero as the network extrapolates beyond the range of its training data set. They point out the need to constrain the gains so that the resulting neural net can be used to extrapolate beyond the range of the data with greater confidence. The Aspen Target product deals with this problem by calculating a model confidence index (MCI) on-line. If the MCI indicates that the neural network prediction is unreliable, the neural net nonlinear map is gradually turned off and the model calculation relies on the linear portion {**A**,**B**,**C**} only. Another feature of this modelling algorithm is the use of extended Kalman filters (EKF) to correct for model-plant mismatch and unmeasured disturbances [31]. The EKF provides a bias and gain correction to the model on-line. This function replaces the constant output error feedback scheme typically employed in MPC practice.

A novel feature of the identification algorithm is that the dynamic model is built with filters and the filter states are used to predict the output variables. Due to the simplistic filter structure, each input variable has its own set of state variables, making the **A** matrix block-diagonal. This treatment assumes that each state variable is only affected by one input variable, i.e. the inputs are decoupled. For the typical case where input variables are coupled, the algorithm could generate state variables that are linearly dependent or collinear. In other words, the resulting model would not be a minimal realisation. Nevertheless, the use of a PLS algorithm makes the estimation of the **C** matrix well-conditioned. The iteration between the estimation of **A**, **B** and **C** matrices will likely eliminate the initial error in estimating the process time constants.

Process nonlinearity is added to the model with concern for model validity using the model confidence index. When the model is used for extrapolation, only the linear portion of the model is used. The use of EKF for output error feedback is interesting; the benefit of this treatment is yet to be demonstrated.

Turner and Guiver of Aspen Technology point out the need to constrain the gains so that the resulting neural net can be used to extrapolate beyond the range of the data with more confidence.[2]

1.3.1.2 Input-output models

The MVC algorithm and the Process Perfecter use input-output models. To simplify the system identification task, both products use a static nonlinear model superimposed upon a linear dynamic model.

Martin *et al.* [18] and later Piche *et al.* [19] describe the details of the Process Perfecter modelling approach. Their presentation is in single-input-single-output form, but the concept is applicable to multi-input-multi-output models. It is assumed that the process input and output can be decomposed into a steady-state portion which obeys a nonlinear static model and a deviation portion that follows a dynamic model. For any input u_k and output y_k, the deviation variables are calculated as follows:

$$\delta u_k = u_k - u_s \tag{11}$$

$$\delta y_k = y_k - y_s \tag{12}$$

where u_s and y_s are the steady-state values for the input and output, respectively and follow a rather general nonlinear relation:

$$y_s = h_s(u_s) \tag{13}$$

The deviation variables follow a second-order linear dynamic relation:

$$\delta y_k = \sum_{i=1}^{2} a_i \delta y_{k-i} + b_i \delta u_{k-i} \tag{14}$$

The identification of the linear dynamic model is based on plant test data from pulse tests, while the nonlinear static model is a neural network built from historical data. It is believed that the historical data contain rich steady-state information and plant testing is needed only for the dynamic submodel. Bounds are enforced on the model gains in order to improve the quality of the neural network for control applications.

The use of the composite model in the control step can be described as follows. Based on the desired output target y_s^d, a nonlinear optimisation program calculates the best input and output values u_s^d and y_s^f using the nonlinear static model. During the dynamic controller calculation, the nonlinear static gain is approximated by a linear interpolation of the initial and final steady-state gains,

$$K_s(u_k) = K_s^i + \frac{K_s^f - K_s^i}{u_s^f - u_s^i}\delta u_k \tag{15}$$

where u_s^i and u_s^f are the current and the next steady-state values for the input, respectively, and

$$K_s^i = \frac{dy_s}{du_s}|_{u_s^i} \tag{16}$$

$$K_s^f = \frac{dy_s}{du_s}|_{u_s^f} \tag{17}$$

which are evaluated using the static nonlinear model. Bounds on K_s^i and K_s^f can be applied. Substituting the approximate gain (15) into the linear submodel yields

$$\delta y_k = \sum_{i=1}^{2} a_i \delta y_{k-i} + \bar{b}_i \delta u_{k-i} + g_i \delta u_{k-i}^2 \tag{18}$$

where

$$\bar{b}_i = \frac{b_i K_s^i (1 - \sum_{j=1}^n a_j)}{\sum_{j=1}^n b_j} \tag{19}$$

$$g_i = \frac{b_i (1 - \sum_{j=1}^n a_j)}{\sum_{j=1}^n b_j} \frac{K_s^f - K_s^i}{u_s^f - u_s^i} \tag{20}$$

The purpose of this approximation is to reduce computational complexity during the control calculation.

It can be seen that the steady-state target values are calculated from a nonlinear static model, whereas the dynamic control moves are calculated based on the quadratic model in (18). However, the quadratic model coefficients (i.e. the local gain) change from one control execution to the next, simply because they are rescaled to match the local gain of the static nonlinear model. This approximation strategy can be interpreted as a successive linearisation at the initial and final states followed by a linear interpolation of the linearised gains. The interpolation strategy resembles gain scheduling, but the overall model is different from gain scheduling because of the gain rescaling. This model makes the assumption that the process dynamics remain linear over the entire range of operation. Asymmetric dynamics (e.g. different local time constants), as a result, cannot be represented by this model.

1.3.1.3 First-principles models

Since empirical modelling approaches can be unreliable and may require a tremendous amount of experimental data, two of the vendors provide the option to use first-principles models. With the NOVA-NLC product, the user specifies the first-principles model through an open equation editor. The PFC model is defined by writing an appropriate software function. In both cases, model parameters must be estimated from plant data.

Hybrid modelling approaches that combine first-principles knowledge with empirical modelling are also found in the commercial packages. The Process Perfecter uses a combination of first-principles models in conjunction with empirical models [20]. The first-principles models can be steady-state balance equations, a nonlinear function of physical variables that generates another physically meaning variable, such as production rate, or simply gain directions to validate empirical models.

1.3.2 Output feedback

In the face of unmeasured disturbances and model errors, some form of feedback is required to remove the steady-state offset. As discussed earlier, the most common method for incorporating feedback into MPC algorithms involves comparing the measured and predicted process outputs [2]. The difference between the two is added to future output predictions to bias them in the direction of the measured output. This can be interpreted as assuming that an unmeasured step disturbance enters at the process output and remains constant for all future time. For the case of a linear model and no active constraints, Rawlings *et al.* [21] have shown that this form of feedback leads to offset-free control. As can be seen in Table 1.2, all five NMPC algorithms described here provide the constant output feedback option.

When the process has a pure integrator, the constant output disturbance assumption will no longer lead to offset-free control. For this case it is common to assume that an integrating disturbance with a constant ramp rate has entered at the output [2]. The PFC, Aspen Target, NOVA-NLC and Process Pefecter algorithms provide this feedback option.

It is well known from linear control theory that additional knowledge about unmeasured disturbances can be exploited to provide better feedback by designing a Kalman filter [22]. Muske and Rawlings demonstrate how this can be accomplished in the context of MPC [23]. It is interesting to note that the Aspen Target algorithm provides an option for output feedback based on a nonlinear generalisation of the Kalman filter known as the extended Kalman filter (EKF) [24]. Aspen Target uses the EKF to estimate both a bias and a feedback gain. It is also possible to implement an EKF within the NOVA-NLC product, although the user is required to enter the appropriate equations manually.

1.3.3 Steady-state optimisation

The PFC, Aspen Target, MVC and Process Perfecter controllers split the control calculation into a local steady-state optimisation followed by a dynamic optimisation. Optimal steady-state targets are computed for each input and output;

these are then passed to a dynamic optimisation to compute the optimal input sequence required to move toward these targets. From Table 1.2 it can be seen that these calculations involve optimising a quadratic objective that includes input and output contributions. The exception is the NOVA-NLC controller that performs the dynamic and steady-state optimisations simultaneously.

1.3.4 Dynamic optimisation

At the dynamic optimisation level, an MPC controller must compute a set of MV adjustments that will drive the process to the steady-state operating point without violating constraints. All of the algorithms described here use a form of the objective given in (4a).

The PFC controller includes only the process input and output terms in the dynamic objective and uses constant weight matrices ($\mathbf{Q}_j = \mathbf{Q}$, $\mathbf{T}_j = 0$, $\mathbf{R}_j = \mathbf{R}$, $\mathbf{S}_j = 0$, $q = 2$). The Aspen Target and MVC products include all four terms with constant weights ($\mathbf{Q}_j = \mathbf{Q}, \mathbf{T}_j = \mathbf{T}, \mathbf{R}_j = \mathbf{R}, \mathbf{S}_j = 0, q = 2$). The NOVA-NLC product adds to this the option of one norms ($\mathbf{Q}_j = \mathbf{Q}, \mathbf{T}_j = \mathbf{T}, \mathbf{R}_j = \mathbf{R}, \mathbf{S}_j = \mathbf{S}, q = 1, 2$).

Instead of using a reference trajectory, the Process Perfecter product ($\mathbf{T}_j = \mathbf{T}, \mathbf{R}_j = 0, \mathbf{S}_j = 0, q = 2$) uses a dynamic objective with trajectory weighting that makes $\mathbf{Q}_j$ gradually increase over the horizon $\mathbf{P}$. With this type of weighting, control errors at the beginning of the horizon are less important than those towards the end of the horizon, thus allowing a smoother control action.

1.3.5 Constraint formulations

There are basically two types of constraints used in industrial MPC technology: hard and soft [2]. Hard constraints are those which should never be violated. Soft constraints allow the possibility of a violation; the magnitude of the violation is generally subjected to a quadratic penalty in the objective function.

All of the NMPC algorithms described here allow hard input maximum, minimum and rate of change constraints to be defined. These are generally defined so as to keep the lower level MV controllers in a controllable range, and to prevent violent movement of the MVs at any single control execution. The PFC algorithm also accommodates maximum and minimum input acceleration constraints which are useful in mechanical servo control applications.

The Aspen Target, MVC, NOVA-NLC and Process Perfecter algorithms perform rigourous optimisations subject to the hard input constraints. The PFC algorithm, however, enforces input hard constraints only after performing an unconstrained optimisation. This is accomplished by clipping input values that exceed the input constraints. It should be noted that this method does not, in general, result in an optimal solution in the sense of satisfying the Karush-Kuhn-Tucker (KKT) conditions for optimality.

All control products enforce output constraints as part of the dynamic optimisation, as shown in (4a). The Aspen Target, NOVA-NLC and Process Perfecter products allow options for both hard and soft output constraints. The PFC product allows only hard output constraints, while the MVC product allows only soft output constraints. The exclusive use of hard output constraints is generally avoided in MPC technology because a disturbance can cause such a controller to lose feasibility.

The Process Perfecter product applies soft constraints by using a frustum method, as depicted in Figure 1.3(a). Compared to the typical zone formulation as shown in Figure 1.3(b), the frustum permits a larger control error in the beginning of the horizon than in the end, but no error is allowed outside the frustum. At the end of the horizon the frustum can have a nonzero zone, instead of merging to a single line, which is determined based on the accuracy of the process model to allow for model errors.

1.3.6 Output trajectories

Industrial MPC controllers use four basic options to specify future CV behaviour; a setpoint, zone, reference trajectory or funnel [2]. All of the NMPC controllers described here provide the option to drive the CVs to a fixed setpoint, with deviations on both sides penalised in the objective function. In practice this type of specification is very aggressive and may lead to very large input adjustments, unless the controller is detuned in some fashion. This is particularly important when the model differs significantly from the true process. For this reason all of the controllers provide some way to detune the controller using move suppression, a reference trajectory or time-dependent weights.

All of the controllers also provide a CV zone control option, designed to keep the CV within a zone defined by upper and lower boundaries. A simple way to implement zone control is to define soft output constraints at the upper and lower boundaries.

The PFC, Aspen Target, MVC and NOVA-NLC algorithms provide a CV reference trajectory option, in which the CV is required to follow a smooth path from its current value to the setpoint. Typically a first- or second-order path is defined using an operator-entered closed-loop time constant. In the limit of a zero time constant the reference trajectory reverts back to a pure setpoint; for this case, however, the controller would be sensitive to model mismatch unless some other strategy such as move suppression is also being used. In general, as the reference trajectory time constant increases, the controller is able to tolerate larger model mismatch.

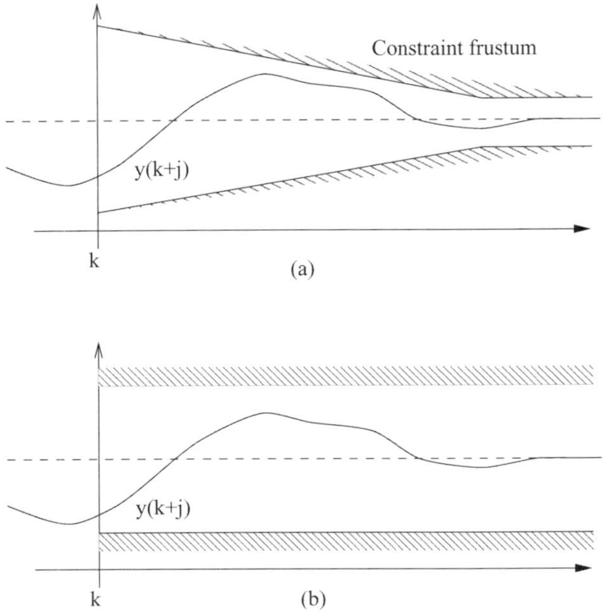

Figure 1.3 Soft constraint formulations: (a) a frustum; (b) a rectangular zone

1.3.7 Output horizon and input parameterisation

Industrial MPC controllers generally evaluate future CV behaviour over a finite set of future time intervals called the prediction horizon [2]. This finite output horizon formulation is used by all of the algorithms discussed in this chapter. The length of the horizon P is a basic tuning parameter for these controllers, and is generally set long enough to capture the steady-state effects of all computed future MV moves. This is an approximation of the infinite horizon solution for closed-loop stability discussed earlier, and may explain why none of the industrial NMPC algorithms considered here includes a terminal state constraint.

The PFC and Aspen Target controllers allow the option to simplify the calculation by considering only a subset of future points called coincidence points, so named because the desired and predicted future outputs are required to coincide at these points. A separate set of coincidence points can be defined for each output, which is useful when one output responds quickly relative to another.

Industrial MPC controllers use three different methods to parameterise the MV profile: a single move, multiple moves and basis functions [2]. The MVC product computes a single future input value; the PFC controller also provides this option. The Aspen Target, NOVA-NLC and Process Perfecter controllers can compute a sequence of future moves spread over a finite control horizon. The length of the control horizon M is another basic tuning parameter for these controllers. Better control performance is obtained as M increases, at the expense of additional computation.

The PFC controller parameterises the input function using a set of polynomial basis functions. This allows a relatively complex input profile to be specified over a large (potentially infinite) control horizon, using a small number of unknown parameters. This may provide an advantage when controlling nonlinear systems. Choosing the family of basis functions establishes many of the features of the computed input profile; this is one way to ensure a smooth input signal, for example. If a polynomial basis is chosen then the order can be selected so as to follow a polynomial setpoint signal with no lag. This feature is important for mechanical servo control applications.

1.3.8 Solution methods

The PFC controller performs an unconstrained optimisation using a nonlinear least-squares algorithm. The solution can be computed very rapidly, allowing the controller to be used for short sample time applications such as missile tracking. Some performance loss may be expected, however, since input constraints are enforced by clipping.

The Aspen Target product uses a multi-step Newton-type algorithm developed by De Oliveira and Biegler [25, 26] and makes use of analytical model derivatives. Due to the sparseness of the state-space model in Aspen Target, the derivative computation is straightforward. The Newton algorithm makes use of the QPKWIK solver, which has the advantage that intermediate solutions, although not optimal, are guaranteed feasible. This permits early termination of the optimisation algorithm if the optimum is not found within the sampling time. Aspen Target uses the same QPKWIK engine for local steady-state optimisation and dynamic MV calculation.

The MVC and Process Perfecter products use a generalised reduced gradient (GRG) code called GRG2 developed by Lasdon and Warren [27]. The NOVA-NLC product uses the NOVA optimisation package, a proprietary mixed complementarity nonlinear programming code developed by DOT Products.

1.4 NMPC application examples

The past five years have seen rapid progress in the development and application of industrial NMPC technology. Nearly 90 industrial applications of NMPC technology were reported in an earlier survey [8]. The number of actual NMPC applications is likely to be significantly larger since only a handful of vendors were included in that survey and only those applications known to the vendors were reported. In fact end-users have applied these products to other problems or have developed their own NMPC algorithms [28]. The number of applications can be expected to grow rapidly in the near future.

While MPC applications are concentrated in refining [2], reported NMPC applications cover a much broader range of application areas. Areas with the largest number of reported NMPC applications include chemicals, polymers and air and gas processing. It has been observed that the size and scope of NMPC applications are typically much smaller than those of linear MPC applications [29]. This is likely due to the computational complexity of NMPC algorithms.

In the following subsections we describe specific applications of each of the NMPC products from Table 1.1.

1.4.1 PFC: application to batch reactors

The PFC product is distinct from other nonlinear MPC implementations in several ways: its SISO configuration, use of reference trajectories, use of coincidence points and use of a clipped nonlinear least squares solver. PFC is also distinct in terms of its wide variety of applications, including those for military systems, automobiles and, most notably, batch chemical reactors. Preuss *et al.* [30] report one such application of PFC for temperature control of batch reactors.

The batch reactor described in Preuss *et al.* [30] is a continuously stirred tank in which exothermic reactions take place. During normal operation the coolant passes through a heat exchanger to remove heat, but during the start-up phase the coolant is warmed in a separate heater. The PFC control configuration features a master PFC reactor temperature control that drives two other PFC algorithms in a slave loop. The slave loop PFC algorithms control the heater and heat exchanger through a split-range, cascade control arrangement.

Since batch reactors involve frequent start-up and shut-down, process nonlinearity is unavoidable, making traditional PID control unsatisfactory. To deal with process nonlinearity, Preuss *et al.* [30] use a linearised first-order model for the heat exchanger, but the gain and the time constant vary with the manipulated variable. The process gain varies with the manipulated variable in a linear relationship, making the overall input-output relation quadratic. This implementation is similar to the one used in the Process Perfecter.

The PFC uses basis functions for the control move parameterisation instead of the typical multiple move approach. Since batch reactors typically ramp the temperature up and down linearly, or hold it constant during the reaction phase, it is convenient for PFC to use two basis functions: a step function and a linear ramp function. This choice makes the computation very fast. The reported control results in Preuss *et al.* [30] show that very small tracking errors are achieved, while PID controllers cannot be tuned satisfactorily in all batch phases.

1.4.2 Aspen Target: application to a pulverised coal fired boiler

Zhao *et al.* [31] reported an application of the Aspen Target controller to a pulverised coal fired boiler control in a 200 MW power plant. The objectives are to

(i) improve boiler efficiency, (ii) reduce NOx emission and (iii) reduce loss of ignition. The process consists of pulverisers for crushing the coal to improve firing, boilers and a turbine. The coal burners are swirled type low-NOx burners with a boiler steam capacity of 650 tons/h. The overall boiler process includes 6 controlled variables, 11 manipulated variables and 16 disturbance variables. Controlled variables for this application are:

- NOx emission level (2)
- CO emission level (2)
- flue gas temperatures (2).

Control is achieved by manipulating the following variables:

- total air flow
- total air dampers (2)
- secondary air dampers (2)
- OFA dampers (2).

Measured disturbances are:

- oxygen concentration (2)
- combustion chamber temperature (2)
- net generated power
- combustion chamber pressure
- flue gas fan motor power (2)
- total air fan motor power (2)
- energy produced in steam (2)
- mills data.

From industrial practice it is known that boilers powered by pulverised coal can be difficult to control, especially regarding lowering NOx emissions. During coal combustion, moisture and oxygen are believed to dominate the formation of NOx and the model relation is nonlinear. Using the Aspen Target controller, Zhao *et al.* [31] report that the Aspen Target controller was able to reduce NOx emission by 15–25 per cent while increasing boiler efficiency by 0.1–0.3 per cent and decreasing loss of ignition by 2 per cent. The control system was shown to be robust under mill changes and rapid load changes within the operating limits of 135–200 MW.

1.4.3 MVC: application to an ammonia plant

Poe and Munsif [32] describe an application of the MVC product to control a plant producing 1450 tons/day of ammonia using natural gas and air as feedstocks. Key factors used to justify this application include:

- dynamic market supply and demand effects on natural gas price and product prices
- capacity and throughput limitations
- variations in gas feedstock rates, quality and composition
- environmental limitations.

The basic objective of the MVC controller is to maximise a profit function computed by subtracting natural gas feed and fuel gas costs from ammonia product, carbon dioxide and steam export revenues. To achieve this the controller uses an overall economic optimisation module to compute optimal steady-state targets for the plant, which are sent to seven separate dynamic controllers:

- hydraulic (steam and pressure balance) module
- primary reformer furnace temperature control module
- primary reformer riser temperature balance control module
- secondary reformer module
- shift/methanator module
- carbon dioxide removal module
- ammonia converter module.

Poe and Munsif [32] describe the control modules in some detail; here we focus only on the ammonia converter module. The ammonia converter is a standard Kellogg 'quenchconverter' design, consisting of three nearly adiabatic catalyst beds, between which fresh feed is introduced to cool the reaction products. The converter control module manipulates the feed flow to the first bed as well as the quench flows to all three beds. This is done in order to maintain the three bed inlet temperatures at their optimal steady-state targets. Output constraints considered by the control include bed outlet temperatures and quench flow valve positions. Feedforward control is provided for changes in feed flowrate, temperature and pressure, hydrogen/nitrogen ratio and inert composition.

Figure 1.4 shows results for the second converter bed; results for the other beds were similar. The MVC controller was able to significantly reduce temperature variations at the bed inlet and outlet, allowing the average reaction temperature to be increased without violating the bed outlet constraint. Overall the plant's net fuel consumption was lowered by 1.8 per cent, while the net production of ammonia increased by 0.7 per cent.

1.4.4 NOVA-NLC: application to a polymerisation process

In an interesting paper presented at the Chemical Process Control VI conference, Young *et al.* [28] discuss the development of the NOVA-NLC algorithm and describe an application to a polymerisation process. The control technology was developed over a five year period by ExxonMobil control engineers in collaboration with two academic groups. After patenting the technology [33] it

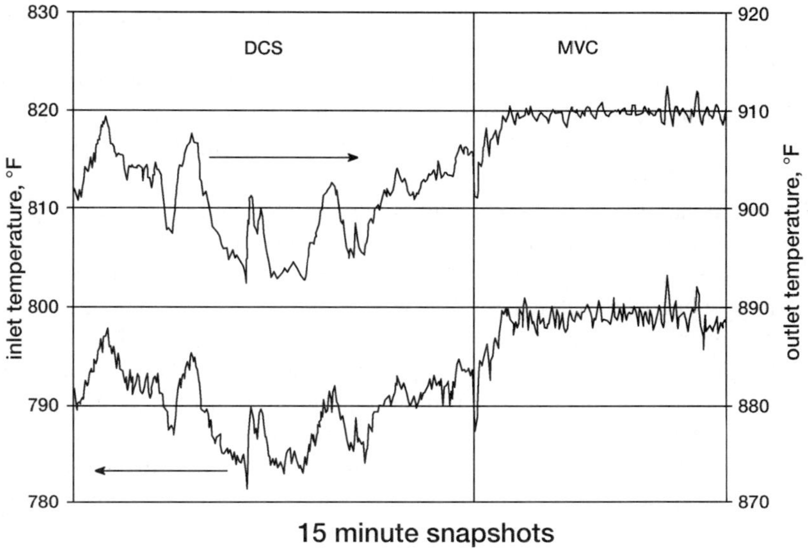

Figure 1.4 MVC results: ammonia converter bed 2 temperatures

was licensed to Dynamic Optimization Technology (DOT) Products for commercialisation.

The nonlinear control development work was originally motivated by the inability to control a new polymerisation process that was brought on-line in 1990. While details of the process are proprietary, they were unable to achieve satisfactory control performance causing methods that had been successful on similar units. The ExxonMobil team contracted with two academic groups to develop an initial model and control algorithm, which were both refined based on further internal development efforts.

The development team chose to use a first-princples nonlinear model because they believed this would lead to the lowest life-cycle cost for their polymer process applications. They specifically ruled out the use of multiple linear models and gain scheduling because of the additional overhead required to develop and maintain the resulting controllers. A second critical choice made by the development team involves the use of a reference trajectory to specify desired closed-loop behaviour. Operating personnel strongly prefer to see the polymer properties follow a constant smooth trajectory during grade transitions. This cannot be achieved using input move-suppression unless the suppression factors are adjusted continuously along the trajectory. Using a second-order reference trajectory allows this specification to be met easily with no re-tuning. The first application was completed successfully in 1994, and to date ExxonMobil has implemented this technology on more than ten other reactor systems.

Young *et al.* [28] describe an application of the NOVA-NLC controller to a polymerisation process with two reactors in series. Each reactor has independent feed and cooling systems. The model includes mass balances for seven species in multiple phases, as well as energy balances around the reactors and cooling systems. The overall model has approximately 120 differential algebraic equations (DAEs) describing roughly 50 states. The controller executes once every 6 min, and the control algorithm requires two to three minutes to compute a solution running on DEC Alpha System 1000 processor.

Controlled variables for this application are polymer melt viscosity and polymer comonomer incorporation within each reactor. Manipulated variables are setpoints in the distributed control system that affect the addition of comonomer and a transfer agent entering each reactor. The application is designed to maintain desired polymer properties at a particular operating point, as well as to follow a specified trajectory during transitions from one product grade to another. Figures 1.5–1.8 illustrate typical performance achieved during a grade transition. Note that Young *et al.* [28] omitted the plot axis scales because this information is proprietary. Overall this represents a significant improvement in performance relative to previous controllers. Typical benefits include cutting the polymer transition time in half.

1.4.5 Process Perfecter: application to a polypropylene process

Demoro *et al.* [20] reported a successful application of the Process Perfecter product to a polypropylene process. This polymer production process is extremely nonlinear with complex reactions and involves frequent grade changes.

Demoro *et al.* [20] also compare the results of a linear MPC with those of a NMPC on the same process, demonstrating the additional benefits of NMPC for this type of process.

The polypropylene process carries out a Ziegler-Natta polymerisation in a bulk liquid-phase (slurry) loop reactor. The control and optimisation objectives are:

- minimise product variability
- minimise grade transition time
- maximise production rate.

The following three controlled variables were selected:

- production rate, calculated from a heat balance
- solids in slurry, calculated from the density of the slurry
- melt flow, measured via lab analysis using melt flow rheometry about every 4 h.

The three selected manipulated variables are:

- catalyst flow
- feed flow of monomers
- modifier concentration, cascaded with the modifier flow controller.

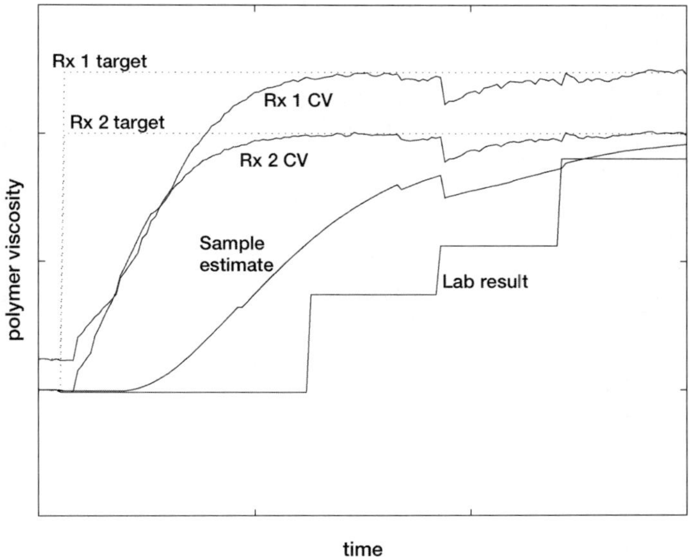

Figure 1.5 NOVA-NLC results: polymer melt viscosity response during grade transition (reprinted with the permission of the CACHE Corporation)

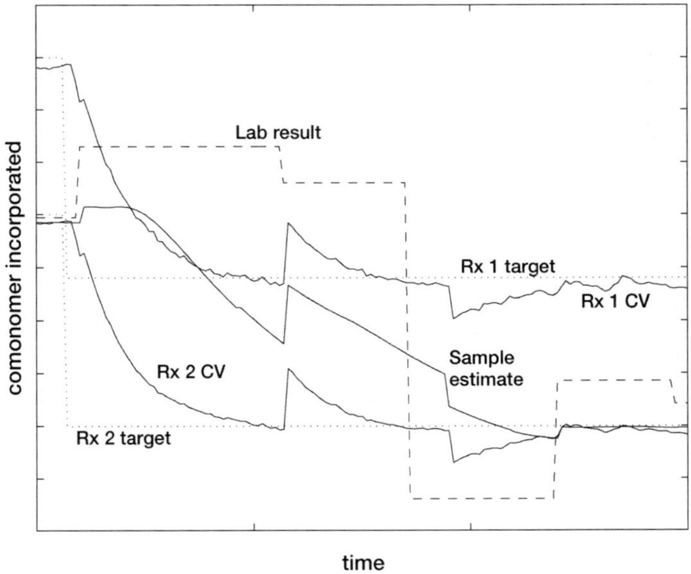

Figure 1.6 NOVA-NLC results: polymer comonomer incorporation response during grade transition (reprinted with the permission of the CACHE Corporation)

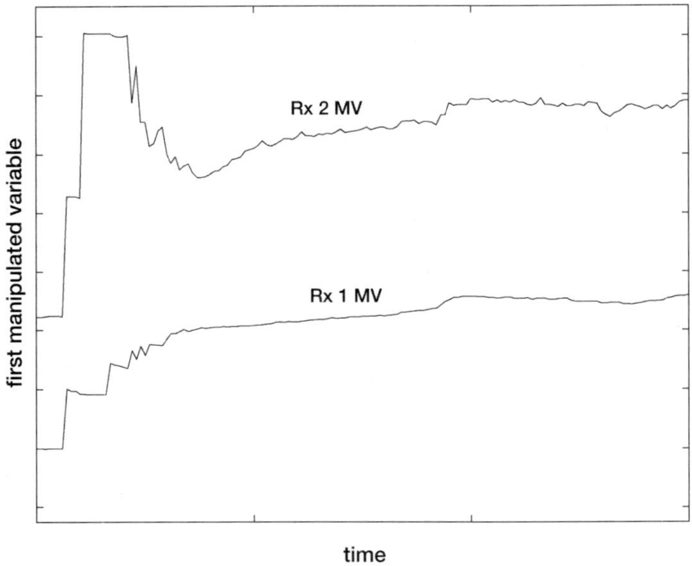

Figure 1.7 NOVA-NLC results: transfer agent command signal during grade transition (reprinted with the permission of the CACHE Corporation)

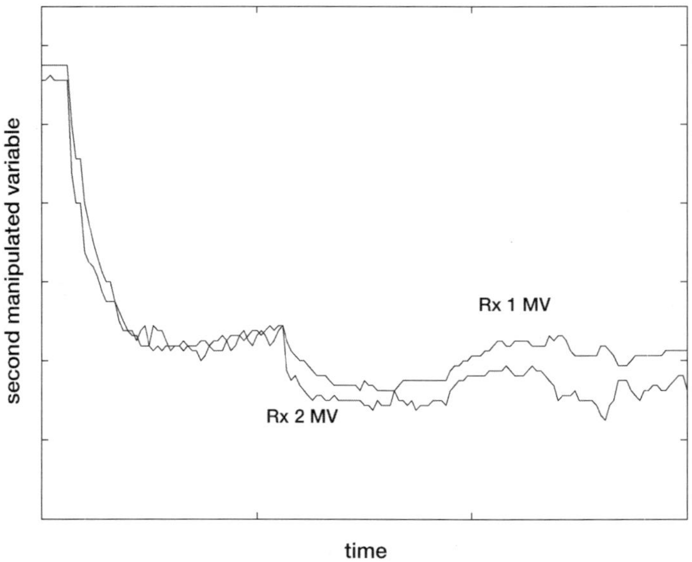

Figure 1.8 NOVA-NLC results: comonomer command signal during grade transition (reprinted with the permission of the CACHE Corporation)

The following two feedforward disturbance variables are used in the NMPC:

- inert concentration in the monomer feed
- reactor temperature.

Because none of the CVs is directly measured on-line, soft sensors were built using either first principles or neural networks from process data. The reactor production rate was calculated with steady-state mass and energy balance models. The amount of solids in slurry was calculated from the density. The melt flow rate was estimated using a neural network, with input variables being either original sensor measurements or transformed based on physical understanding of the process, for instance, production rate and component hold-up. Since several sensors are analytical measurements which are prone to gross errors, a sensor validation module screens the data before it is used in optimisation and control.

The static nonlinear process model was then built from historical data using neural networks, which also estimated the melt flow rate from lab test data. This model was integrated with a dynamic linear submodel using the Process Perfecter software. To determine the benefits of using NMPC for this process, a linear model was also built for the process to configure a linear MPC.

Demoro *et al.* [20] reported three control experiments on the polypropylene process. The first one used the linear model, the second used the nonlinear model, and the third used the nonlinear model in conjunction with an upper level nonlinear optimiser. In each experiment the polymer melt flow rate was changed from 30 to 35. It was observed that the linear MPC was unable to accomplish the transition within 30 min, since the process is highly nonlinear and the controller was detuned to achieve stability. The nonlinear MPC algorithm was able to accomplish the transition very quickly. In the third experiment where an optimiser was used, production rate was maximised while the grade transition was accomplished within the process constraints.

1.5 Future needs for NMPC technology development

The following sections highlight aspects of NMPC technology where significant issues remain to be resolved.

1.5.1 Model development

There is no systematic approach for building nonlinear dynamic models for NMPC. In the case of empirical approaches, guidelines for plant tests are needed to build a reliable model. This is important because even more test data will be required to develop an empirical nonlinear model than an empirical linear model. In the case of rigourous physical models, links must be provided to allow such models to be downloaded from standard dynamic modelling software packages. The ideal product would allow models to be developed using a combination of physics and test data. It is sometimes the case, for example, that one can write a mass balance that holds rigourously for a process unit, whereas the energy and momentum balances have a high degree of uncertainty. Another useful option would be to allow models for different parts of a plant to be combined in a seamless manner, perhaps by dragging icons together in a desktop design environment.

1.5.2 Output feedback

Most current NMPC implementations use a constant bias to correct the model based on current measurements. However, this approach suffers from several known limitations even for linear systems. Muske and Rawlings [23] recognised that this is equivalent to assuming that a step disturbance enters at the output of the process. They went on to show that a wider class of disturbance models can be implemented using a standard Kalman filter. If it is known that process disturbances enter through a process input, for example, much better performance can be achieved by utilising an input disturbance model.

For nonlinear systems it seems reasonable to expect that similar benefits can be achieved by implementing an explicit disturbance model using EKF. Guidelines for disturbance model design are needed, however, to ensure that the resulting augmented system is detectable and that it allows for offset-free control.

Moving horizon estimation (MHE) may eventually provide the best solution to the output feedback problem. In this method the state estimate is found through an on-line optimisation, allowing state constraints to be incorporated directly into the calculation. The most sensible way to formulate the MHE problem is still an open question; relevant issues are discussed in References 3 and 34.

1.5.3 Optimisation methods

Speed and the assurance of a reliable solution in real time are major limiting factors in existing applications. Research continues on how best to formulate and solve the NMPC problem, with notable recent results reported by Tenny *et al.* [35] and Findeisen *et al.* [36]. The method described by Tenny *et al.* has the advantage that it builds on their interior-point QP solution for the linear MPC problem [5] and it provides a feasible solution in the event that the calculation times out. The method described by Findeisen *et al.* [36] is interesting because it uses a continuous differential algebraic equation (DAE) model as a starting point. These two methods have yet to be compared directly on a meaningful industrial scale problem.

1.5.4 User interface

Simpler and more powerful user interfaces will be required in order to fully exploit the potential of NMPC technology. At a recent conference one speaker likened the NMPC products sold today to the MPC products using linear models that were marketed in the mid 1980s. The packages have a lot of rough edges, but they get the job done. There is much that can be done, however, to hide unnecessary complexity from the user. In principle, the control design interface for an NMPC product should be no more complex than that for an MPC product using linear models.

1.5.5 Justification of NMPC

Design guidelines are needed to indicate when NMPC may provide significant benefits relative to simpler methods. This is especially important for NMPC technology because model development is so expensive. Benchmarks on a array of industrial processes will be required in order to develop sensible design guidelines. Only one such activity has been reported to date [20].

1.5.6 Other issues

Other issues raised in the survey of linear MPC technology [2] may prove to be just as important for NMPC technology. These issues include using sequential optimisations to implement prioritised constraints, automatically screening out ill-conditioned subprocesses, simplifying the tuning effort and designing the controller so that it tolerates faults more easily.

1.6 Conclusions

We can draw the following conclusions.

- The past five years have seen rapid progress in the development and application of NMPC algorithms with a wide range of industrial applications.
- The algorithms reported here differ in the simplifications used to generate a tractable control calculation; all of them, however, are based on adding a nonlinear model to a proven NMPC formulation.
- None of the currently available NMPC algorithms includes the terminal state constraints or infinite prediction horizon required by control theory for nominal stability; instead they rely implicitly upon setting the prediction horizon long enough to effectively approximate an infinite horizon.
- The three most significant obstacles to NMPC applications are: nonlinear model development, state estimation and rapid, reliable solution of the control algorithm in real time.

- Future needs for NMPC technology include development of nonlinear model identification and nonlinear estimation methods, reliable numerical solution techniques, and better guidelines for justifying NMPC applications.

1.7 References

1 RAWLINGS, J.B.: 'Tutorial overview of model predictive control', *IEEE Control Syst. Mag.*, 2000, **20** (3), pp. 38–52

2 QIN, S.J., and BADGWELL, T.A.: 'An overview of industrial model predictive control technology', in KANTOR, J.C., GARCIA, C.E., and CARNAHAN, B. (Eds). Fifth international conference on *Chemical process control*, AIChE and CACHE, 1997, pp. 232–56

3 ALLGÖWER, F., BADGWELL, T.A., QIN S.J., RAWLINGS, J.B., and WRIGHT, S.J.: 'Nonlinear predictive control and moving horizon estimation – an introductory overview', in FRANK, P.M., (Ed.): 'Advances in control: highlights of ECC'99' (Springer, 1999), pp. 391–449

4 MAYNE, D.Q., RAWLINGS, J.B., RAO, C.V., and SCOKAERT, P.O.M.: 'Constrained model predictive control: stability and optimality', *Automatica*, 2000, **36**, pp. 789–814

5 RAO, C.V., WRIGHT, S.J., and RAWLINGS, J.B.: 'Application of interior-point methods to model predictive control', *J. Optim. Theory Appl.*, 1998, **99**, pp. 723–57

6 RICHALET, J., RAULT, A., TESTUD, J.L., and PAPON, J.: 'Model predictive heuristic control: applications to industrial processes', *Automatica*, 1978, **14**, pp. 413–28

7 PRETT, D.M., and GILLETTE, R.D.: 'Optimization and constrained multi-variable control of a catalytic cracking unit'. Proceedings of Joint Automatic Control Conference, 1980

8 QIN, S.J., and BADGWELL, T.A.: 'An overview of nonlinear model predictive control applications', in ALLGÖWER, F., and ZHENG, A. (Eds): 'Nonlinear model predictive control' (Birkhauser Verlag, 2000), pp. 369–92

9 LEE, E.B., and MARKUS, L.: 'Foundations of optimal control theory' (John Wiley & Sons, New York, 1967)

10 SCOKAERT, P.O.M., MAYNE, D.Q., and RAWLINGS, J.B.: 'Suboptimal model predictive control (feasibility implies stability)', *IEEE Trans. Autom. Control*, 1999, **44** (3), pp. 648–54

11 KALMAN, R.E.: 'Contributions to the theory of optimal control', *Bull. Soc. Math. Mex.*, 1960, **5**, pp. 102–19

12 BELLMAN, R.E., and DREYFUS, S.E.: 'Applied dynamic programming' (Princeton University Press, Princeton, New Jersey, 1962)

13 KEERTHI, S.S., and GILBERT, E.G.: 'Optimal infinite-horizon feedback laws for a general class of constrained discrete-time systems: stability and moving horizon approximations', *J. Optim. Theory Appl.*, 1988, **57** (2), pp. 265–93

14 MICHALSKA, H., and MAYNE, D.Q.: 'Robust receding horizon control of constrained nonlinear systems', *IEEE Trans. Autom. Control*, 1993, **38** (11), pp. 1623–33

15 CHEN, H., and ALLGÖWER, F.: 'A quasi-infinite horizon nonlinear model predictive control scheme with guaranteed stability', *Automatica*, 1998, **34** (10), pp. 1205–18
16 SENTONI, G.B., BIEGLER, L.T., GUIVER, J.B., and ZHAO, H.: 'State-space nonlinear process modelling: identification and universality', *AIChE J.*, 1998, **44** (10), pp. 2229–39
17 ZHAO, H., GUIVER, J.P., and SENTONI, G.B.: 'An identification approach to nonlinear state space model for industrial multivariable model predictive control'. Proceedings of 1998 American Control Conference, Philadelphia, PA, USA, 1998
18 MARTIN, G., BOE, E., PICHE, S., KEELER, J., TIMMER, D., GERULES, M., and HAVENER, J.: 'Method and apparatus for dynamic and steady state modeling over a desired path between two end points'. US Patent 5933345, 1999
19 PICHE, S., SAYYAR-RODSARI, B., JOHNSON, D., and GERULES, M.: 'Nonlinear model predictive control using neural networks', *IEEE Control Syst. Mag.*, 2000, **20** (3), pp. 53–62
20 DEMORO, E., AXELRUD, C., JOHNSTON, D., and MARTIN, G.: 'Neural network modelling and control of polypropylene process'. Society of Plastics Engineers International Conference, Houston, TX, 1997
21 RAWLINGS, J.B., MEADOWS, E.S., and MUSKE, K.R.: 'Nonlinear model predictive control: a tutorial and survey'. ADCHEM '94 Proceedings, Kyoto, Japan, 1994
22 KALMAN, R.E., and BUCY, R.S.: 'New results in linear filtering and prediction theory', *Trans. ASME, J. Basic Eng.*, 1961, **83** (1), pp. 95–108
23 MUSKE, K.E., and RAWLINGS, J.B.: 'Model predictive control with linear models', *AIChE J.*: 1993, **39** (2), pp. 262–87
24 RAMIREZ, W.F.: 'Process control and identification' (Academic Press, New York, NY, 1994)
25 DE OLIVEIRA, N.M.C., and BIEGLER, L.T.: 'Constraint handling and stability properties of model-predictive control', *AIChE J.*, 1994, **40** (7), pp. 1138–55
26 DE OLIVEIRA, N.M.C., and BIEGLER, L.T.: 'An extension of Newton-type algorithms for nonlinear process control', *Automatica*, 1995, **31**, pp. 281–6
27 LASDON, L.S., and WARREN, A.D.: 'GRG2 user's guide'. Technical Report, Department of Computer and Information Science, Cleveland State University, Cleveland, OH, 1986
28 YOUNG, R.E., BARTUSIAK, R.D., and FONTAINE, R.W.: 'Evolution of an industrial nonlinear model predictive controller'. Preprints of Chemical Process Control – CPC VI, Tucson, AZ, USA, January 2001, CACHE, pp. 399–410
29 MARTIN, G., and JOHNSTON, D.: 'Continuous model-based optimization'. Hydrocarbon Processing's Process Optimization Conference, Houston, TX, 1998
30 PREUSS, K., LE LANN, M.-V., RICHALET, J., CABASSUD, M., and CASAMATTA, G.: 'Thermal control of chemical batch reactors with predictive functional control', *Journal A*, 1998, **39** (4), pp. 13–20
31 ZHAO, H., GUIVER J., NEELAKANTAN, R., and BIEGLER, L.T.: 'A nonlinear industrial model predictive controller using integrated pls and neural state space model'. 14th IFAC Triennial World Congress, Beijing, PR China, 1999

32 POE. W., and MUNSIF. H.: 'Benefits of advanced process control and economic optimization to petrochemical processes'. Hydrocarbon Processing's Process Optimization Conference, Houston, TX, 1998
33 BARTUSIAK, R.D., and FONTAINE, R.W.: 'Feedback method for controlling nonlinear processes'. US Patent 5682309, 1997
34 RAO, C.V., and RAWLINGS, J.B.: 'Nonlinear moving horizon state estimation', in ALLGÖWER, F., and ZHENG, A. (Eds): 'Nonlinear model predictive control' (Birkhauser Verlag, 2000), pp. 45–69
35 TENNY, M.J., RAWLINGS, J.B., and BINDLISH, R.: 'Feasible real-time nonlinear model predictive control'. Preprints of Chemical Process Control–CPC VI, Tucson, AZ, USA, January 2001, CACHE, pp. 439–43
36 FINDEISEN, R., ALLGÖWER, F., DIEHL, M., BOCK, G., SCHLODER, J., and NAGY, Z.: 'Efficient nonlinear model predictive control'. Preprints of Chemical Process Control – CPC VI, Tucson, AZ, January 2001. CACHE, pp. 454–60
37 BERKOWITZ, P., and PAPADOPOULOS, M.: 'Multivariable process control method and apparatus'. US Patent 5396416, 1995
38 MVC3.0 User Manual. Continental Controls, Inc. Product Literature, 1995
39 BERKOWITZ, P., PAPADOPOULOS, M., COLWELL, L. and MORAN, M.: 'Multivariable process control method and apparatus'. US Patent 5488561, 1996
40 KEELER, J., MARTIN, G., BOE, G., PICHE, S., MATHUR, U., and JOHNSTON, D.: 'The process perfecter: the next step in multivariable control and optimization'. Technical Report, Pavilion Technologies, Inc., Austin, TX, 1996

1.8 Notes

[1]This product was formerly known as NeuCOP II

[2]TURNER, P., and GUIVER, J.: 'Neural network APC. Fact or fantasy'. Submitted to *Control Solutions*, 2000

Chapter 2

Nonlinear model predictive control: issues and applications

Robert S. Parker, Edward P. Gatzke, Radhakrishnan Mahadevan, Edward S. Meadows and Francis J. Doyle III

Abstract

The nonlinear model predictive control (NMPC) algorithm is a powerful control technique with many open issues for research. This chapter highlights a few of these issues through a series of process and biosystems case studies. Control using nonlinear models can be further complicated when working with distributed parameter systems. An emulsion polymerisation process examines these challenges. Efficient solution techniques for NMPC problems are necessary when solution time or constraints are important. A fed-batch bioreactor is used to examine a computationally efficient NMPC algorithm based on differential flatness. For systems where incomplete information is available, the estimation analogue of NMPC, nonlinear moving horizon estimation, can be incorporated into the algorithm. This is demonstrated on the Van der Vusse reactor using multiple linear models. In the absence of a fundamental process description, nonlinear input-output models can be used to characterise the process. As an example, an analytical solution to the 2-norm NMPC problem for SISO systems modelled by Volterra-Laguerre systems is implemented on a continuous bioreactor. Paths for future research are also identified.

2.1 Introduction

Model predictive control (MPC) is a popular control algorithm which solves an optimisation problem on-line at each time step. For problems adequately described by linear models, the linear MPC algorithm is an efficient algorithm which incorporates inherent multivariable and constraint handling capabilities. In some cases, however, the selection of the desired operating range coupled with possibly nonlinear process dynamics can degrade performance and potentially destabilise the closed-loop system (e.g. high-purity distillation and systems with an extremum). Nonlinear MPC (NMPC) can alleviate this performance degradation, while retaining the multivariable and constraint handling benefits of the MPC algorithm. Significant effort has been focused on NMPC, as evidenced by a collection of sessions and plenaries over the past ten years [1–8].

There are still many open issues in the synthesis, analysis and application of NMPC controllers. Few comprehensive tools are available for the development of high-fidelity nonlinear dynamic models. Once the model has been constructed (either a fundamental model or via input-output model identification) issues in problem solution arise. The resulting nonlinear programming problems are typically nonconvex and constrained. This can lead to infeasibilities for output constrained problems, as well as implementation issues when sample times are shorter than the time required to solve the optimisation problem. Computational efficiency is also necessary when large-scale systems or complex fundamental models are integrated as part of the solution algorithm at each iteration. Uncertainty can play a major role in controller performance; robustness issues, such as robust stability or performance, sensitivity to plant-model mismatch, and the ability to estimate disturbances from process data, are important in controller implementation. Additionally, NMPC does not guarantee that a globally optimal result will be returned; local minima can trap some search algorithms yielding suboptimal performance. This chapter examines a few of the open issues in NMPC: (i) exploitation of nonlinear model structure; (ii) algorithm computational efficiency; (iii) control of systems described by population balance models; and (iv) disturbance estimation using process residuals.

2.2 Exploiting model structure

2.2.1 Motivation

Strongly nonlinear systems, especially those displaying even-order nonlinear behaviour, cannot be controlled at the optimum by a linear integrating controller [9]. Optimal control via local linearisation of the nonlinear process at each sample time has been studied,[1] although full state measurement and an accurate process model were required. Full state measurement is rarely available in practice, thereby limiting the utility of this technique. NMPC implementation requires a nonlinear model, but the construction of a high-fidelity nonlinear model is nontrivial. When a fundamental model is unavailable, or when a highly detailed nonlinear model is prohibitively complex to implement, the use of input-output models is a possible alternative. One popular structure for nonlinear modelling and controller design is the Volterra series [10–12]. Note that Volterra models are not able to capture arbitrary nonlinearities (e.g. output multiplicity or chaotic dynamics) but can approximate chemical process behaviours such as asymmetric response to symmetric input changes and input multiplicity. This model structure is useful for approximating process responses which do not depend 'too strongly' on past input values [13]. This section will concentrate on modelling and controller synthesis for the class of systems which can be represented adequately by second-order Volterra series (or related input-output) models.

2.2.2 Model identification

A second-order Volterra series model can be decomposed as:

$$\begin{aligned} y(k) &= h_0 + \mathscr{D}(k) + \mathscr{O}(k) + \mathscr{L}(k) \\ \mathscr{L}(k) &= \sum_{i=1}^{M} h_1(i)u(k-i) \\ \mathscr{D}(k) &= \sum_{i=1}^{M} h_2(i,j)u^2(k-i) \\ \mathscr{O}(k) &= 2\sum_{i=1}^{M}\sum_{j=1}^{i-1} h_2(i,j)u(k-i)u(k-j) \end{aligned} \tag{1}$$

where the linear, second-order diagonal and off-diagonal component can be assumed symmetric. Parsimonious methods for Volterra model identification have been presented previously [12, 14, 15]. Tailored input sequences excite specific contributions, thereby resulting in improved coefficient estimates when compared to cross-correlation. Two common complaints regarding the use of Volterra models, however, come from the highly parameterised structure of the model and the significant variation in the coefficient estimates when identified from noisy process data. One filtering technique involves a projection of the Volterra series model onto the Laguerre basis, resulting in a Volterra-Laguerre model [11]. By using the orthogonal basis functions, the high-order Volterra model can be significantly reduced in order. The second-order Volterra-Laguerre model resulting from the projection of a Volterra series model is given (in state-space form) by [16]:

$$l(k+1) = A(\alpha)l(k) + B(a)u(k) \tag{2}$$

$$\hat{y}(k) = C^T l(k) + l^T(k)Dl(k) \tag{3}$$

Here the linear state equations are dependent only on the user-defined Laguerre pole, $(0 < \alpha \leq 1)$. The C and D matrices are calculated via least squares from the identified Volterra kernels. Use of this method to identify an input-output model from a continuous-stirred bioreactor process [17, 18] was shown in Reference 12.

2.2.3 Controller synthesis

The objective function used in this NMPC formulation is the squared 2-norm given by:

$$\min_{\Delta \mathbf{U}(k)} \left\| \Gamma_y[\mathcal{Y}(k+1|k) - \mathcal{R}(k+1)] \right\|_2^2 + \left\| \Gamma_u \Delta \mathcal{U}(k) \right\|_2^2. \tag{4}$$

Here the formalism of solving for $\Delta\mathcal{U}$ rather than absolute $\mathcal{U}$ is utilised, with variable weight. This optimisation is solved at each sample time over a prediction horizon of length p, for a series of m moves which minimise the objective. Given a model of the form in (2) and (3), an analytical solution to (4) has been constructed for the case where $m = 1$ and the process is single-input single-output (SISO) [12]. Limits on input-output dimension and m were imposed by the inability to analytically solve a third-order vector or matrix polynomial, respectively. Recently, Zheng [19–21] has proposed a computationally efficient method for solving NMPC problems where the first move is calculated using the (potentially constrained) nonlinear problem, and all subsequent manipulated variable moves are determined via a linear MPC controller. Although the goal of the current work is not necessarily computational efficiency, the concept of approximating input move changes after the first move is appealing because it allows the analytical solution of the multiple move problem ($m > 1$).

The approximation of future move calculations by a linear controller allows a restructuring of the NMPC objective function:

$$\min_{\Delta \mathbf{U}(k)} \left\| \Gamma_y[\mathcal{R}(k+1) - \mathcal{Y}_N(k+1|k) - \mathcal{Y}_L(k+1|k)] \right\|_2^2 + \left\| \Gamma_u \Delta \mathcal{U}(k) \right\|_2^2 \tag{5}$$

The predicted future effect of the first move on the output is nonlinear (given by $\mathcal{Y}_N$, and governed by (2) and (3)), while moves after the first affect the output linearly ($\mathcal{Y}_L$). In this way, the linear problem can be formulated as a solution to a modified reference signal, $\mathcal{R}(k+1) - \mathcal{Y}_N(k+1|k)$, which is dependent on $\Delta u(k|k)$. Hence, the linear controller problem can be solved as an explicit function of $\Delta u(k|k)$, the first move. The linear controller model is given by:

$$x_L(k+i|k) = \left\{ \begin{array}{cc} 0 & i = 1 \\ \sum_{j=1}^{i} \bar{A}_{m-j-1} B \Delta u(k+j|k) & 2 \le i \le m-1 \\ A^{i-m+1} \sum_{j=1}^{i} \bar{A}_{m-j-1} B \Delta u(k+j|k) & i \ge m \end{array} \right\} = G \Delta \mathbf{U}_K(k|k)$$

$$y_L(k+i|k) = [C^T + 2x_{lin}^T D] x_L(k+i|k) = H x_L(k+i|k) \tag{6}$$

Here $\bar{A}_i = (A^{i-1} + A^{i-2} + \cdots + I)$, $\Delta\mathcal{U}_L(k|k)$ is the set of $m-1$ linear manipulated variable moves, and x_{lin} is the value of the state around which the linearisation is developed. The matrices G and H are constant at any sample time, and are introduced for convenience. For the second-order systems studied here, a minimum of $2x_{lin}$ sets must be used (one for each side of the process optimum). The determination of x_{lin} could be performed dynamically on-line throughout the controller computation, but an analytic solution is not available for this case. Hence, the static controller method described by Zheng [19, 21] is utilised in this work. The linear problem can therefore be solved as:

$$\Delta \mathbf{U}_L(k|k) = (G^T H^T \Gamma_{yL}^T \Gamma_{yL} HG + \Gamma_{uL}^T \Gamma_{uL})^{-1} G^T H^T \Gamma_{yL}^T \Gamma_{yL} (\mathscr{R}(k+1) - \mathscr{Y}_N(k+1|k)) \tag{7}$$

The weighting matrices $\Gamma_{yL} = \Gamma_y(2:p,\ 2:p)$ and $\Gamma_{uL} = \Gamma_y(2:m,\ 2:m)$ are the respective dimensions consistent with the linear problem solution. Equation (5) can be solved analytically by substituting (7) into the objective function and solving explicitly for $\Delta u(k|k)$. The fourth-order polynomial which results is a function of $l(k)$ (the current nonlinear model state), $u(k-1)$ (the immediate past input) and $\Delta u(k|k)$:

$$(\mathscr{R}(k+1) - \mathscr{Y}_N(k+1|k))^T [\mathscr{I}^T \Gamma_y^T \Gamma_y \mathscr{I} + \mathscr{K}^T \Gamma_{uL}^T \Gamma_{uL} \mathscr{K}] \mathscr{R}(k+1) - \mathscr{Y}_N(k+1|k) - \Delta u^2(k|k) \Gamma_u''(1,1) \tag{8}$$

where:

$$\mathscr{K} = (G^T H^T \Gamma_{yL}^T \Gamma_{yL} HG + \Gamma_{uL}^T \Gamma_{uL})^{-1} G^T H^T \Gamma_{yL}^T \Gamma_{yL} \tag{9}$$

$$\mathscr{I} = I_{p-1 \times p-1} - HG\mathscr{K} \tag{10}$$

$$\mathscr{Y}_N(k+i|k) = \varepsilon_0 + \varepsilon_1 \Delta u(k|k) + \varepsilon_2 \Delta u^2(k|k) \tag{11}$$

$$\varepsilon_2(k+i|k) = B^T \bar{A}_i^T D \bar{A}_i B \tag{12}$$

$$\varepsilon_1(k+i|k) = C^T \bar{A}_i B + 2l^t(k)(A^i)^T D \bar{A}_i B + 2\varepsilon_2(k+i|k) u(k-1) \tag{13}$$

$$\varepsilon_0(k+i|k) = C^T (A^i) l(k) + l^T(k)(A^i)^T D (A^i) l(k) + \varepsilon_1(k+i|k) u(k-1) \tag{14}$$

$$\bar{A}_i = (A^{i-1} + \cdots + I_{3\times 3}) \tag{15}$$

The vectors E_2, E_1 and E_0 are the stacked elements of ε_2, ε_1 and ε_0, respectively, for all $i = 1, \ldots, p$. Once defined, the above polynomial can be differentiated with respect to $\Delta u(k|k)$, and set equal to 0, resulting in the equation:

$$0 = \xi_3 \Delta u^3(k|k) + \xi_2 \Delta u^2(k|k) + \xi_1 \Delta u(k|k) + \xi_0 \tag{16}$$

where:

$$\xi_3 = 2E_2^T\Gamma_{yN}^T\Gamma_{yN}E_2 \tag{17}$$

$$\xi_2 = 3E_1^t\Gamma_{yN}^T\Gamma_{yN}E_2 \tag{18}$$

$$\xi_1 = 2E_2^T\Gamma_{yN}^T\Gamma_{yN}E_0 + E_1^T\Gamma_{yN}^T\Gamma_{yN}E_1 - 2E_2^T\Gamma_{yN}^T\Gamma_{yN}\mathscr{R}(k+1) +_u^2 (1,1) \tag{19}$$

$$\xi_0 = E_1^t\Gamma_{yN}^T\Gamma_{yN}E_0 - E_1^T\Gamma_{yN}^T\Gamma_{yN}\mathscr{R}(k+1) \tag{20}$$

$$\Gamma_{yN} = \begin{bmatrix} \Gamma_y(1,1) & 0 \\ 0 & \mathbf{I}^T\Gamma_{yL}^T\Gamma_{yL}\mathbf{I} + \mathscr{K}^T\Gamma_{uL}^T\Gamma_{uL}\mathscr{K} \end{bmatrix} \tag{21}$$

Provided that the ξ_is are real, an analytic solution to (16) exists [22], and the optimal value for $\Delta u(k|k)$ can be calculated as developed for the $m = 1$ case in Reference 12.

2.2.4 *Application: a continuous bioreactor*

A growth model for Klebsiella pneumoniae on glucose in a continuous-flow bioreactor is given in Baloo and Ramkrishna [17, 18]. Biomass exit concentration (g/l) is the output of interest, and dilution rate (h^{-1}) is the manipulated input. The steady-state dilution rate for this process is 0.97 h^{-1}, which corresponds to a biomass exit concentration of 0.2373 g/l. A second-order Volterra-Laguerre model for this process using a sample time of 5 min was identified in Reference 12. NMPC for $m = 1$ in the presence of input magnitude constraints was also presented, and was demonstrated to be superior to the asymptotic solution as well as gradient-based NMPC techniques under certain conditions due to the second-order nature of the steady-state locus [12].

The extension to $m > 1$ for reference tracking is presented in this work. A reachable setpoint was selected, meaning that the reference intersected the steady-state locus (as opposed to the unreachable case where the reference is above the maximum achievable steady-state value). The magnitude constraints on this problem were $0.2 \le u(k) \le 1.1\,h^{-1}$, such that the cells did not starve or wash out of the reactor, respectively. The analytical NMPC controller was then tested for a -0.05 g/l change in the reference. Cell biomass profiles are shown in Figure 2.1. Clearly, the increase in the move horizon, m, improves controller performance in terms of a sum-squared error (SSE) metric (dashed is superior to solid response by nine per cent). The addition of a recursive least-squares (RLS) algorithm to this problem is straightforward [16, 23, 24], even for the $m > 1$ case as the $\mathscr{I}$ and $\mathscr{K}$ matrices are updated at each sample time. A significant performance improvement was expected, as a more accurate model on the low dilution rate side of the optimum would improve model prediction accuracy. Settling time in the RLS case (shown by the solid line) is significantly reduced, and performance improves by 35 per cent over the $m = 1$ case without RLS (14 per cent improvement over

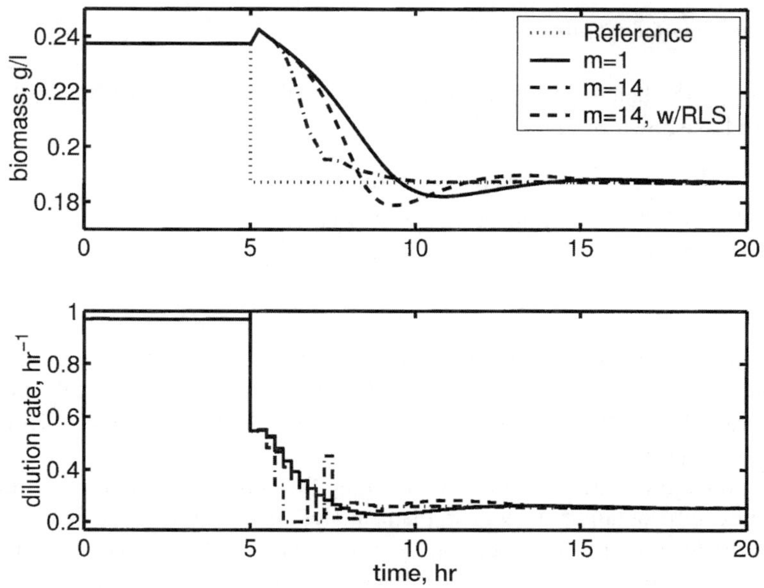

Figure 2.1 *Response to a -0.05 g/L reference change at $t = 5$ h. Tuning parameters are: $p = 16$; $\Gamma_u(1,1) = 0, \Gamma_{ul} = I_{m-1\times m-1}$, sample time = 15 min. Top: cell biomass concentration; bottom: dilution rate*

$m = 14$ without RLS). As expected, the improvement in performance is a strong function of the weighting functions.

2.3 Efficient dynamic optimisation using differential flatness

2.3.1 Motivation

Central to the NMPC scheme is the solution of the dynamic optimisation to compute the next set of control moves for a given objective. This has been discussed in Reference 3 and more recently in Reference 1. In this section, an approach based on the concept of differential flatness is presented for the formulation of the NMPC problem for a fed-batch case as a lower dimensional nonlinear program (NLP). This technique effectively combines the solution of the model differential equations and the dynamic optimisation into a single NLP problem. The problem is cast as a continuous time formulation with feedback being incorporated by resetting the initial conditions as the measurement is available.

2.3.2 Problem formulation

Consider a fed-batch problem where a terminal objective function has to be optimised as shown in (22):

$$\begin{aligned} \max_{x(t),u(t)} \quad & \Phi(x,u)|_{t=t_f} \\ \text{subject to} \quad & \dot{x} = f(x,u) \\ & 0 \geq c(x,u) \forall t \in [t_0, t_f] \\ & x|_{t=t_0=x_0} \end{aligned} \tag{22}$$

Here Φ is the terminal objective function, f is the nonlinear function describing the process, c are the constraints, t_f, t_f, are the initial and the final time and $x \in \mathbf{R}^{n_x}, x_0 \in \mathbf{R}^{n_x}, u \in \mathbf{R}^{n_u}$ are the states, initial conditions and the manipulated variables, respectively.

This problem can be solved in an open-loop fashion and the optimal profiles for the input can be obtained. However, plant-model mismatch, parametric uncertainty and disturbances cause this open-loop profile to be suboptimal. Thus, feedback in the form of either estimating the disturbances/model parameters and updating the model or changing the initial conditions of the model followed by solving the dynamic optimisation problem is essential to obtain the updated input profile. In the proposed approach, the optimisation problem is solved the instant a measurement becomes available after modifying the initial conditions.

The optimisation problem can be recast as a lower dimensional NLP if the system in (22) admits the property of differential flatness. A system of ordinary differential equations is differentially flat if all the states and inputs can be represented in terms of the flat outputs (ξ) and its derivatives as shown below. The flat outputs themselves are functions of the states, inputs and finite derivatives of the inputs (for additional details on this subject, the reader is referred to the introductory articles [25, 26]):

$$\begin{aligned} x &= \chi(\xi, \dot{\xi}, \ldots, \xi^{\eta}) \\ u &= \mathbf{Y}(\xi, \dot{\xi}, \ldots, \xi^{\kappa}) \end{aligned} \tag{23}$$

The equation above can be used to transform the optimisation problem in (22) as shown below:

$$\begin{aligned} \max_{\xi(t)} \quad & \tilde{\Phi}(\xi, \dot{\xi}, \ldots, \xi^{\kappa}) \\ \text{subject to} \quad & 0 \geq \tilde{c}(\xi, \dot{\xi}, \ldots, \xi^{\kappa}) \\ & \chi(\xi, \dot{\xi}, \ldots, \xi^{\eta})|_{t=t_0} = x_0 \end{aligned} \tag{24}$$

where $\tilde{\Phi}$ is the new objective function, $\tilde{c}$ are the transformed constraints and $\xi, \dot{\xi}, \ldots, \xi^{\kappa} \in \mathbf{R}^{n_y}$ are the flat outputs and their derivatives. In the transformed problem, the system dynamic constraints are not required as the nonlinear system is fully represented by the flat outputs and their derivatives. Thus the new problem is to find the optimal profile of the flat outputs and use (23) to obtain the state and the input profiles. Now this problem is recast as an NLP by representing the flat outputs as shown below:

$$\xi(t) = \Omega_0(t) + \sum_{i=1}^{P} \Theta_i(t) Z_i \tag{25}$$

where $\Theta_i(t)$ are the basis functions, Z_i are the corresponding coefficients, $\Omega_0(t)$ is the function satisfying the initial conditions and P is the number of basis elements chosen for each flat output. Thus the total number of coefficients is $n_u P$. The constraints are imposed at a finite number of points in time. The transformed NLP problem is given below:

$$\begin{aligned} \max_{\mathbf{Z}} \quad & \hat{\Phi}(\mathbf{Z}, t)|_{t=t_f} \\ \text{subject to} \quad & 0 \geq \tilde{c}(\mathbf{Z}, t_j) j = 1 \ldots M \\ & t_j = t_0 + (j-1)\Delta T \\ & \Delta T = (t_f - t_o)/(M-1) \end{aligned} \tag{26}$$

where $\hat{\Phi}$ is the transformed terminal objective function, $\hat{c}$ are the constraints, M is the number of points at which the constraints are imposed and ΔT is the time interval between these points. Φ, $\tilde{\Phi}$ and $\hat{\Phi}$ have the same dimension. The total number of constraints ($\hat{c}$) after the transformation to an NLP increases M-fold. One of the advantages of this approach is that the number of free parameters is relatively small compared to existing optimisation approaches. This approach is illustrated in the example in the following section.

2.3.3 Application: biomass optimisation

The objective of this bioreactor application is to maximise the amount of biomass formed at the end of the batch. A nonlinear bioreactor model including constraints is given by:

$$\begin{aligned} \dot{x}_1 &= \mu(x)x_1 - ux_1/x_3 \\ \dot{x}_2 &= u(S_f - x_2)/x_3 - \mu(x)x_1/Y_{X/S} \\ \dot{x}_3 &= u \\ \mu(x) &= \mu_m x_2/(K_m + x_2) \end{aligned} \tag{27}$$

$$0 \leq u \leq 10 \mathrm{l\,h}^{-1} \text{ and } x_3 \leq 10 \mathrm{l} \tag{28}$$

The states in the nonlinear model admit the following relation:

$$\begin{aligned} x_3(x_1 + Y_{X/S}(x_2 - S)) &= C \\ C &= x_3^o(x_1^o + Y_{X/S}(x_2^o - S_f)) \end{aligned} \tag{29}$$

where C is a constant and the x_i^os denote the initial conditions. One of the states (biomass x_1) is eliminated using the equation above, to obtain a two-state model as shown in (30) (here y_1, y_2 refer to x_2, x_3 respectively):

$$\begin{aligned}\dot{y}_1 &= u(S_f - y_1)/y_2 - \frac{\mu_m y_1 (C/y_2 - Y_{X/S}(y_1 - S_f))}{(K_m + y_1)Y_{X/S}} \\ \dot{y}_2 &= u\end{aligned} \tag{30}$$

The nonlinear model of the bioreactor in (30) is differentially flat with the flat output $\xi = Y_{X/S}(y_1 - S_f)y_2$. The representation of the states and inputs in terms of the flat output and its derivatives are not shown here for the sake of brevity. The details relating the flat outputs and the states and inputs appear in Mahadevan *et al.* [27]. The optimisation problem to be solved as a part of NMPC every time a measurement is available is summarised below:

$$\begin{aligned}\max_{u(t)} \quad & \Phi(t_f) = x_1 x_3 = (C - Y_{X/S}S_2(y_1 - S_f))|_{t=t_f} \\ \text{subject to} \quad & 0 \leq u \leq 10\,1/\text{h} \\ & y_2 \leq 10\,1 \\ & y|_{t=t_f} = y_0.\end{aligned} \tag{31}$$

Using the relations between the flat outputs and the states and the inputs, the above problem is transformed to the following NLP (32). Here, the batch time is divided into N equal intervals and the flat output is parameterised by a fifth degree polynomial in each time interval:

$$\begin{aligned}\max_{\mathbf{Z}} \quad & \hat{\Phi}(\mathbf{Z}, t_f) = (C - \xi(\mathbf{Z}))|_{t=t_f} \\ \text{subject to } & 0 \geq u[\xi(Z)] \leq 10\,1\,\text{h}^{-1} \; \forall t \in [t_0, t_f] \\ & y_2[\xi(Z)]|_{t=t_f} = 10\,1.\end{aligned} \tag{32}$$

The flat output and its higher derivatives are restrained to be continuous at the beginning of all the time intervals except the first one, where initial conditions are imposed. Thus we have $3N + 1$ free parameters for optimisation. Also, in order to account for singularities, an additional constraint is imposed to prevent the NLP solver from drifting into singular regions. The results of the simulation for a case where one of the model parameters (μ_m) is time varying is shown in Figure 2.2.

Since the constraints are enforced at a finite number of points in time, minor constraint violations in between these points are observed (see Figure 2.2). Preliminary analysis of the computational times for open-loop optimisation with an arbitrary initial guess indicates the efficiency of this method is superior to methods like the simultaneous approach where all the states and the inputs are parameterised. The computation of the open-loop optimal profiles on a Sun Microsystems Ultra10 with a 333 MHz processor takes approximately 15 s for the proposed approach as compared to approximately 150 s for the simultaneous approach. With a good initial guess, the corresponding times can be decreased to 3 and 50 s, respectively. It should be noted that the symbolic computation of the flat outputs and the relation between these outputs and the states is demanding. As model order increases these calculations represent a potential limitation to this approach.

2.4 Model-based control of population balance systems

2.4.1 Motivation: emulsion polymerisation

Emulsion polymerisation represents an economically important industrial process that produces a wide range of polymer products including paints and surface coatings, glues and adhesives. For products like these that are used in the form of a latex, the final product properties depend upon particle size distribution (PSD) and molecular weight distribution (MWD), not just average values. In recent years, detailed models for PSDs and MWDs have been developed, and computing power has increased to the point where it is feasible to incorporate detailed models in an on-line feedback scheme [28–43]. Coupled with effective feedback, the use of model-based control for emulsion polymerisation offers the promise of reduced variability of latex properties through direct control of particle size and molecular weight distributions.

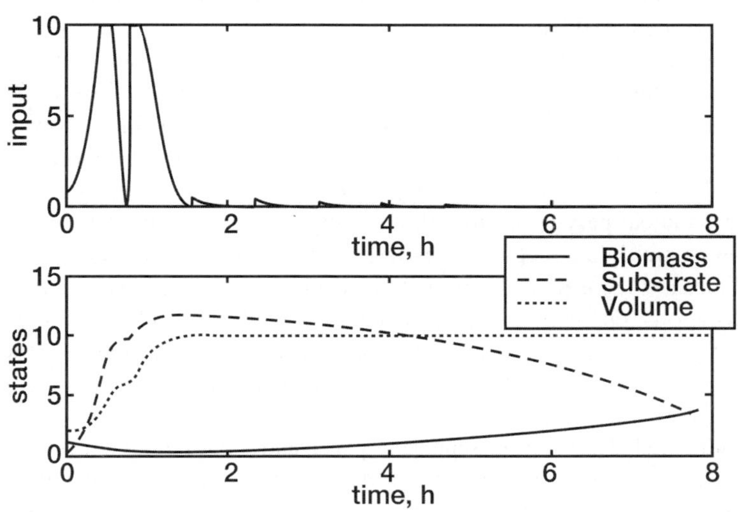

Figure 2.2 Closed-loop optimisation under parametric uncertainty in μ_m. Formulation parameters: $N = 19$, $M = 40$. Top: feed flow rate; bottom: model states (biomass, substrate and volume)

Although control studies and experiments at the University of Delaware have focused on emulsion polymerisation, many of the issues and ideas involved have wider applicability. For example, the basic equations that describe evolution of PSD and MWD are similar to those for crushing and grinding in solids processing, for the formation of atmospheric aerosols, for the growth of cells in bioreactors and for the evolution of animal populations in nature.

2.4.2 Model development

For a large population that can be described by a single variable, e.g. radius of particles in an emulsion, the evolution of the population is given by the following system involving a partial differential equation:

$$\frac{\partial}{\partial t}(Vn) + V\frac{\partial}{\partial r}(Gn) = \mathbf{B}(x, n, r, t) - \mathbf{D}(x, n, r, t) \tag{33}$$

$$\dot{x} = f(x, u, n, t) \tag{34}$$

$$y = g(x, u, n, t) \tag{35}$$

in which $n = n(r, t)$ is the population density, $G = G(x, r, t)$ is the growth rate, V is the reactor liquid volume, and **B** and **D** represent birth and death rates. In (34), x represents state variables in the reaction system, such as chemical species concentrations and temperature, that affect the evolution of the PSD, and can be adequately described using ordinary differential equations. Since emulsion polymerisation reactors may be operated as semi-batch, x may also include reactor volume V (shown explicitly in (33)).

A control vector u has been included in the right-hand side of (34) to reflect our interest in manipulating the system through x to control the PSD n. Available controls for an emulsion polymerisation system may include flow rates of monomers, solvents, surfactant, initiator or chain transfer agent, mixing speed and heat transfer rate to cooling or heating media.

For free-radical polymerisation, it is usually necessary to model the concurrent evolution of multiple populations that are distinguished by the number of radicals per particle. For rapid radical termination within particles, it is sufficient to model two populations, one containing no radicals (n_0) and a second containing a single radical (n_1). In this case, a partial differential equation of the form of (33) would be required for both n_0 and n_1. Details of this so-called zero-one model may be found in Reference 37.

The birth and death terms represent gains and losses within the population. Contributions to these terms are derived from ordinary mass and energy balances for particles and chemical species within the reactor, incorporating our understanding of the polymerisation rate and other chemical reactions occurring in each phase, and accounting for the mass transfer of free radical species between particles and the aqueous media.

As in many complicated polymerisation systems, values of kinetic and thermodynamic parameters may not be available from the literature or may have been reported with widely varying values. Our work includes sensitivity studies to determine which parameters are key to obtaining successful feedback control, and which will be used to guide parameter estimation experiments.

Comprehensive treatment of modelling of emulsion polymerisation reactors may be found in the cited references [29, 38, 40, 44].

2.4.3 Numerical solutions of the population balance equation

Most approaches to numerical solution of (33) rely on discretising the size dimension r and approximating the partial derivatives with respect to r. One of the chief difficulties with these schemes for emulsion polymerisation systems lies in dealing efficiently with particle growth. For example, an efficient discretisation scheme would be more finely discretised where the curvature (i.e. $|\partial^2 n/\partial r^2|$) of the probability density function (PDF) was high. Particle growth shifts peak from the PSD into regions that previously had no particles, suggesting that the discretisation should adapt to accommodate particle growth. Several authors have considered such approaches for problems with moving fronts. A standard reference for moving discretisation is Reference 45.

Our approach uses orthogonal collocation on finite elements [46, 47] (see Figure 2.3) to discretise the particle radius and convert (33) into a set of coupled ordinary differential equations. Preliminary work has been performed using a fixed finite element grid; however, we anticipate extending our methods to moving element boundaries to accommodate a wider range of particle growth.

2.4.4 Approaches to control

Using our model, it is hypothetically possible to implement a model predictive control (MPC) algorithm that will drive the system to the desired final product conditions. We might view this approach as the 'ideal' way to implement model based feedback control; however, several factors complicate implementation of a typical MPC scheme, including finite stopping time and the computational demands of large-scale nonlinear optimisation.

Operation in batch or semi-batch mode means that there is no steady-state value that can be used as a setpoint. This invalidates the standard MPC feedback structure which is designed to estimate the steady-state offset between the model and the process, so an alternate dynamic approach to estimating model mismatch must be used.

To alleviate the twin problems of excessive computational demands and lack of a steady-state operating point, we can adopt a hybrid strategy that combines

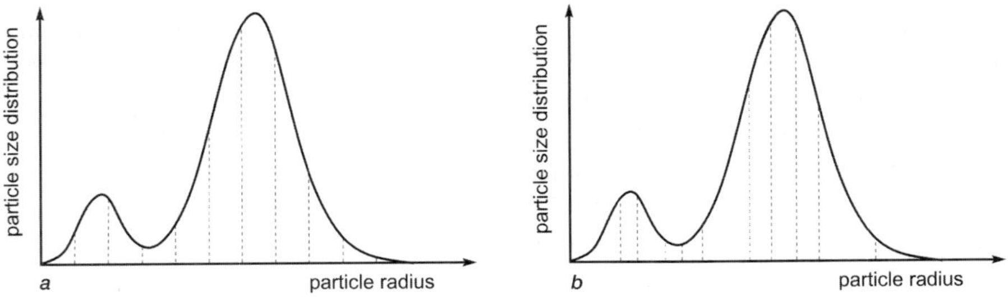

Figure 2.3 Grids for finite element discretisation: (a) uniform; (b) adaptive

nonlinear open-loop optimisation with a closed-loop linear controller. In this method, we compute the desired trajectory infrequently using the full nonlinear model, and taking the current best estimate of model mismatch at the terminal time. This step is computationally demanding and may be performed off-line. A regulatory MPC-based controller can then be implemented using a time-varying linear model obtained from a linearisation about the nominal desired trajectory. An estimate of model mismatch at the final time may be computed at each time step, and if it exceeds a user-specified value, a full nonlinear reoptimisation of the trajectory could be performed.

2.4.5 Measurement and feedback

Probably the most difficult aspect of model-based control of emulsion polymerisation stems from the paucity of measurements. In most industrial scale reactors, measurements are limited to variables such as temperature, level and pressure. Such measurements are useful in trying to maintain the semibatch trajectory close to a prespecified recipe, but are only indirectly related to particle size or molecular weight distributions. A goal of work at the University of Delaware is to implement on-line feedback control of particle size distribution through direct measurements of PSD obtained via capillary hydrodynamic fractionation [48], a method that has only recently been incorporated into commercial instrumentation. In addition to the direct measurement of PSD, our work also includes on-line analysis of monomer species in the reactor head space. This is intended to provide feedback information about the phase distribution of monomers in the system.

Even with these measurements, it is necessary to study the controllability and observability of states in our model in order to construct a satisfactory feedback controller. We anticipate that the fineness of the discretisation for numerical solution of (33) will be determined through this controllability and (especially) observability. In this way, we can view model reduction as essentially equivalent to adopting a coarser discretisation.

2.4.6 Batch polymerisation example

Emulsion polymerisation is often conducted in batch or semibatch reactors. In a typical batch recipe, chemical reagents are mixed with water, and the mixture is agitated and heated to begin the process of polymerisation. The initial mixture consists of water, monomer, surfactant and free radical initiator.

The course of batch emulsion polymerisation can be described as proceeding through three distinct intervals, characterised by the species and phases present in each. In each interval, polymerisation takes place in the monomer-rich interior of polymer particles. Interval I is characterised by the presence of micelles, monomer droplets and growing, surfactant-stabilised polymer particles. During Interval I, micelles become polymer particles as monomer and free radicals diffuse into micelles and cause their growth via polymerisation. As polymer particles grow, they adsorb more surfactant onto their surfaces and the concentration of surfactant in the bulk solution falls. When bulk surfactant concentration drops below the critical micelle concentration (CMC), the continued existence of micelles is no longer thermodynamically favourable and micelles disappear.

The fall of free surfactant levels below the CMC characterises the transition to Interval II. With no more micelles, the mechanism for forming new particles stops and no new particles are generated. However, monomer droplets and aqueous free radicals are still present, and their continuing diffusion into polymer particles results in continued particle growth via polymerisation.

The disappearance of monomer droplets as a separate phase marks the onset of Interval III. During this period, the continuing presence of aqueous free radicals drives up conversion as monomer within particles polymerizes. As monomer is consumed, the character of polymer particles changes, and some of our modelling assumptions may no longer hold.

Some results from a simulation of a typical batch run are shown in Figure 2.4, showing monomer concentration within particles, aqueous surfactant concentration and fractional conversion. In the model described in this work, we considered styrene as the monomer, sodium dodecyl sulfate (SDS) as the surfactant and sodium persulfate as the initiator. The transitions between intervals are clearly indicated. This multi-regime character and the presence of threshold effects in batch emulsion polymerisation present several difficulties in designing model based control. All efficient codes for nonlinear optimisation of smooth functions depend upon local gradient information. This can be highly misleading in the presence of threshold effects; for example, if surfactant addition is a control variable, its effect on particle generation may be zero when surfactant is below the critical micelle concentration (CMC) and very large above it. The effect is very pronounced and may dictate the use of non-gradient-based approaches to optimisation.

An alternate approach might be to implement inferential cascade controllers in which model-based feedback might be used to determine variables such as particle nucleation rate or free surfactant level relative to the CMC. The profiles for these inferential variables would be decided via nonlinear programming, off-line or at greatly reduced sampling frequency.

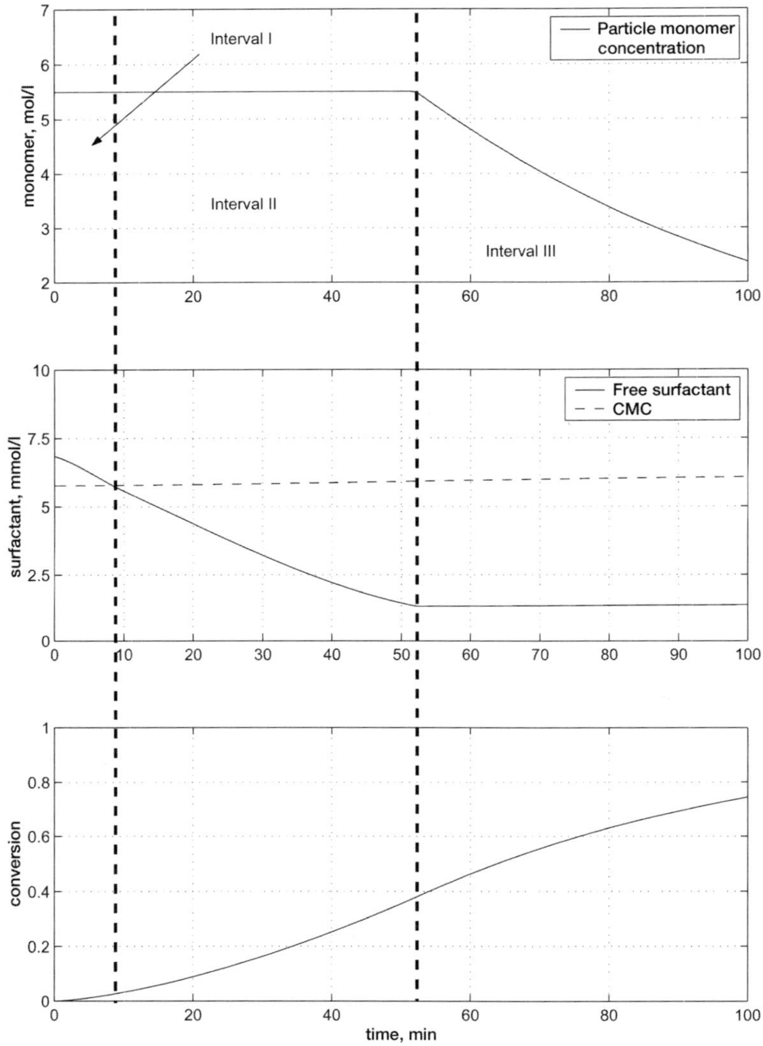

Figure 2.4 A typical batch run showing Intervals I–III

2.5 Disturbance estimation

2.5.1 Motivation

Uncontrollable external influences affect process systems, potentially resulting in adverse changes in product quality. If the disturbance is directly measurable, feedforward control can be used in conjunction with traditional feedback methods to aggressively counteract the disturbance effects. In cases where the necessary additional on-line sensors are not available, process measurements and process models can be used to accurately and effectively estimate unmeasured system disturbances. These estimates can be used to update a process model in a control algorithm, enabling rapid disturbance rejection and minimisation of quality variations.

Moving horizon estimation is a method with many similarities to moving horizon control. In moving horizon control, an optimisation problem is solved at every sample time to minimise the difference between a reference and process model prediction into the future, searching over possible control moves. In moving horizon estimation, an optimisation problem is solved at every sample time to minimise the difference between a limited set of the past process measurements and the estimated measurements from a process model, searching over the possible process disturbances. In both the control and estimation problems, a new measurement is received from the process at each sample time, and the optimisation problem is solved. See References 49–54 for moving horizon estimation examples. An advantage of moving horizon estimation over other estimation methods is its straightforward inclusion of constraints in the optimisation (estimation) problem.

Constraints can be used to represent assumptions about a process and enable the use of multiple models to represent nonlinear systems. In some cases, one can assume that many different disturbances affect a process, but only a limited number of the disturbances affect the system at any one time. This assumption can be implemented in a moving horizon estimation framework as a propositional logic constraint, involving integer variables representing the presence of each individual disturbance [55–58]. The dynamic response of real systems to a disturbance may have various nonlinear characteristics. Potential process nonlinearities include asymmetric responses to symmetric inputs or disturbances, varying characteristic time constants and changing gains. Multiple linear models can be used to capture the nonlinearity of these types of processes. Constraints can be developed such that positive and negative changes in a disturbance are treated as separate disturbances. Small and large disturbance changes can be modelled separately, allowing the estimation problem to optimally select a combination of models to best fit the process measurements. The resulting constrained mixed integer optimisation problem must then be solved on-line.

2.5.2 Estimation formulation

A process model is used to account for known manipulated input moves. The difference between the actual process measurements and the process model are considered the process measurement residuals, y. Let y be the estimated output response to a disturbance Θ. At each sample time, the optimisation problem is solved:

$$\min_{\Theta(i), i=k-H+1,\ldots,k} \|Q(y(i) - \hat{y}(i))\|_1 + \|R\Delta\Theta(i)\|_1 \tag{36}$$

subject to the constraints:

$$\Delta\Theta(i) = \Theta(i) - \Theta(i-1) \tag{37}$$

$$\Theta_j(i) = \theta_{j,1}(i) + \cdots + \theta_{j,n_j}(i) \tag{38}$$

$$\hat{y}_o = \sum_{j=1}^{F} M_{o,j,1}\theta_{j,1} + \cdots + M_{o,j,n_j}\theta_{j,n_j} \quad \forall o \tag{39}$$

$$\sum_{i=1}^{k-H+1} \sum_{n=1}^{n_j} \theta_{j,n}(i) \le Pf_j \quad \forall j \tag{40}$$

$$\sum_{j=1}^{F} f_j \le S \tag{41}$$

$$\theta_{;j,n} \ge 0 \quad \forall j, nf_j \in \{0, 1\} \tag{42}$$

In the objective function, k is the current time, y_i is the process measurement residual vector at time i, $\hat{y}(i)$ is the vector of process model residual estimates at time i, Q is the measurement weighting vector, R is the weighting vector on changes in the parameter estimate, H is the length of the moving horizon, and $\Theta(i)$ is the vector of disturbance estimates at time i. The value $\Delta\Theta(i)$ is the change in a parameter from one time step to another, with Θ_{i-H} as the estimate θ_{i-H+1} from the previous sample time result. An impulse response formulation is used to describe the response of the process model estimate, $\hat{y}(i)$, to changes in the system parameters, Θ. Individual system disturbances (or faults) are described at time i as $\Theta_j(i)$, where j is the index describing distinct disturbances. In this formulation, there may be multiple models for a single model disturbance affecting the model outputs. The individual disturbance estimate Θ_j for disturbance j at time i is the sum of the contributions from each of the n models for that disturbance, where n_j models are used for a given disturbance. Mixing coefficients are not used to select models based on operating regime, as the models are selected to best fit the available data during the optimisation. Assuming that there are o different process residuals available, the output residuals $\hat{y}_o$ are calculated with M_{o,j,n_j} as the impulse response coefficients for model M_{o,j,n_j}. F is the total number of disturbances modelled in the formulation. To handle asymmetric response to a parameter change, two models can be used for each disturbance. The values for all $\theta_{j,n}(i)$ can be constrained to be positive values (≥ 0). This formulation assumes that both large positive and large negative parameter changes will not occur in a parameter over a single window length.

At this point the formulation only includes continuous variables and few constraints. Solving this formulation without additional constraints typically yields an underspecified problem that matches the measurement values with the estimated measurement values exactly. One can make the assumption that only a limited number of disturbances can affect the system during a single horizon. This leads one to use binary decision variables, f_j, to represent whether or not a disturbance has affected the system in the current horizon window. The value P is a large number which ensures that, whenever a disturbance $\theta_{j,n_j}(i)$ is nonzero, f_j switches from 0 to 1. S is the total number of faults that can occur in a horizon window. Because of the binary variables and 1-norm objective function, (36)–(42) represent a mixed integer linear program (MILP). This problem could have been formulated with a 2-norm objective (structurally similar to the NMPC objective in (4)), thereby requiring the solution of a more computationally intensive mixed integer quadratic program (MIQP).

2.5.3 Application: chemical reactor disturbance estimation

A nonlinear system model for the Van der Vusse continuous stirred tank reactor (CSTR) is used to simulate process data. A full description of the Van der Vusse CSTR model may be found in Reference 59 or 60. This system exhibits many significant nonlinear behaviours, including input multiplicity, asymmetric response to symmetric input changes, and transition from minimum- to non-minimum-phase behaviour. A feed stream of feedstock A enters a reactor and reacts to form the desired product, B. The model assumes a first-order reaction for $A \Rightarrow B$ with two competing reactions $B \Rightarrow C$ and $2A \Rightarrow D$. For this example, the following ten potential disturbances were considered: input feed flow rate; input temperature; input concentration; coolant flow rate; coolant temperature; heat transfer coefficient; and biases in the four state measurements. Four step response models were used for each disturbance, corresponding to small and large changes in both the positive and negative directions. In studies where only one disturbance affects the system over the estimation horizon, the estimation routine can distinguish the correct disturbance and provide an accurate estimate, as seen in Figures 2.5 and 2.6. Similar studies (not shown) have demonstrated that the formulation can also yield estimates for multiple faults occurring in an estimation horizon. The disturbance estimates can result in indirect supervisory response or direct control action. Overall, moving horizon estimation using multiple models and logic constraints can result in significant computational complexity, but the application of advanced estimation techniques can also lead to improved controller performance.

2.6 Conclusions

Issues in the development of nonlinear moving horizon control and estimation have been discussed through a series of motivating process case studies. Each of these issues has considered some type of tradeoff, typically between solution accuracy and computational efficiency. Modelling of distributed parameter systems could

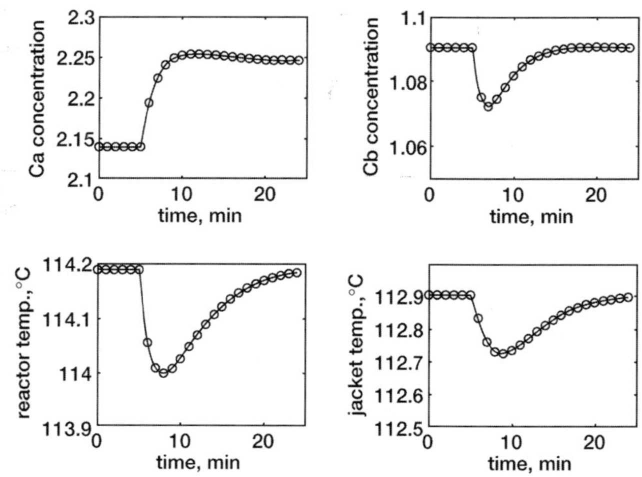

Figure 2.5 *Process measurement residuals for an unmeasured 10 per cent increase in feed flow at time = 5*

lead to the use of more accurate nonlinear process models in the NMPC controller, although the implementation of these models will require further exploration in terms of simulation and solution methods, as well as constraint handling. Less complicated nonlinear modelling approaches, like the multi-model method used in

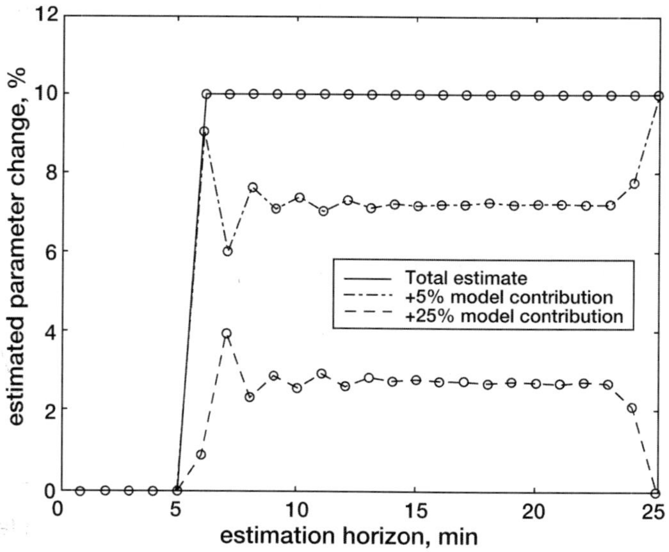

Figure 2.6 *Feed flow rate estimates for the 10 per cent increase, simultaneously using both +5 per cent and +25 per cent models for estimation*

the nonlinear moving horizon estimator, may be easier to implement, but rigorous stability and performance analysis are more complicated. Efficient algorithms for solving NMPC problems posed using differential flatness are being developed, although this approach may not be feasible for high-dimension systems. The approximation of complicated processes using low-order input-output models may sacrifice model accuracy, but clever selection of the model structure may facilitate the solution of the NMPC problem.

2.7 Acknowledgments

Funding for this work has been provided by the National Science Foundation (CTS 9257059, BES-9896061), the Office of Naval Research Young Investigator Program (NOOO-14-96-1-0695) and the University of Delaware Process Control and Monitoring Consortium.

2.8 References

1 ALLGÖWER, F., BADGWELL, T.A., QIN, J.S., RAWLINGS, J.B., and WRIGHT, S.J.: 'Nonlinear predictive control and moving horizon estimation – an introductory overview', in 'Advances in control – highlights of ECC '99' (Springer, London, 1999), pp. 391–449

2 BIEGLER, L.T.: 'Efficient solution of dynamic optimisation and NMPC problems'. International symposium on *Nonlinear model predictive control: assessment and future directions*, Ascona, Switzerland, 1998, pp. 46–63

3 BIEGLER, L.T., and RAWLINGS, J.B.: 'Optimization approaches to nonlinear model predictive control'. ARKUN, Y., and RAY, W. (Eds). Proceedings of the 4th international conference on *Chemical process control*, Padre Island, TX, 1991, pp. 543–71

4 BIEGLER, L.T.: 'Advances in nonlinear programming concepts for process control'. IFAC symposium on *Advanced control of chemical processes*, Banff, Canada, 1997, pp. 587–98

5 LEE, J.H.: 'Modelling and identification for nonlinear predictive control: requirements, current status and future research needs'. International Symposium on *Nonlinear model predictive control: assessment and future directions*, Ascona, Switzerland, 1998, pp. 91–107

6 MAYNE, D.Q.: 'Nonlinear model predictive control: an assessment'. Proceedings of the 5th international conference on *Chemical process control*, Tahoe City, CA, 1996

7 QIN, S.J., and BADGWELL, T.A.: 'An overview of nonlinear MPC applications'. International symposium on *Nonlinear model predictive control: assessment and future directions*, Ascona, Switzerland, 1998, pp. 128–45

8 RAWLINGS, J.B., MEADOWS, E.S., and MUSKE, K.R.: 'Nonlinear model predictive control: a tutorial and survey'. IFAC Symposium on *Advanced control of chemical processes*, Kyoto, Japan, 1994, pp. 203–14
9 MORARI, M.: 'Robust stability of systems with integral control'. Proceedings of the IEEE conference on *Decision and control*, San Antonio, TX, 1983, IEEE Press, pp. 865–9
10 DOYLE III, F.J., OGUNNAIKE, B.A., and PEARSON, R.K.: 'Nonlinear model-based control using second-order Volterra models', *Automatica*, 1995, **31**, pp. 697–714
11 ZHENG, Q., and ZAFRIOU, E.: 'Nonlinear system identification for control using Volterra-Laguerre expansion'. Proceedings of American Control Conference, Seattle, WA, 1995, pp. 2195–9
12 PARKER, R.S., and DOYLE III, F.J.: 'Nonlinear model predictive control of a continuous bioreactor at near-optimum conditions'. Proceedings of American Control Conference, Philadelphia, PA, 1998, pp. 2549–53
13 BOYD, S., and CHUA, L.O.: 'Fading memory and the problem of approximating nonlinear operators with Volterra series', *IEEE Trans. Circuits Syst.*, 1985, **CAS-32** (11), pp. 1150–61
14 DOYLE III, F.J., PARKER, R.S., PEARSON, R.K., and OGUNNAIKE, B.A.: 'Plant-friendly identification of second-order Volterra models'. In Proceedings of European Control Conference, Karlsruhe, Germany, 1999
15 HEEMSTRA, D.G.: 'Practical nonlinear model identification and control implementation'. Master's Thesis, Purdue University, 1996
16 FU, Y., and DUMONT, G.A.: 'An optimum time scale for discrete Laguerre network', *IEEE Trans. Autom. Control*, 1993, **38**, pp. 934–8
17 BALOO, S., and RAMKRISHNA, D.: 'Metabolic regulation in bacterial continuous cultures: I', *Biotech. Bioeng.*, 1991, **38**, pp. 1337–52
18 BALOO, S., and RAMKRISHNA, D.: 'Metabolic regulation in bacterial continuous cultures: II', *Biotech. Bioeng.*, 1991, **38**, pp. 1353–63
19 ZHENG, A.: 'A computationally efficient nonlinear MPC algorithm'. Proceedings of American Control Conference, Albuquerque, NM, 1997, pp. 1623–7
20 ZHENG, A.: 'Nonlinear model predictive control of the Tennessee Eastman process.' Proceedings of American Control Conference, Philadelphia, PA, 1998, pp. 1700–4
21 ZHENG, A., and ALLGÖWER, F.: 'Towards a practical nonlinear predictive control algorithm with guaranteed stability for large-scale systems.' Proceedings of American Control Conference, Philadelphia, PA, 1998, pp. 2534–8
22 TUMA, J.J.: 'Engineering mathematics handbook' (McGraw-Hill, New York, NY, 1987)
23 DUMONT, G.A., FU, Y., and LU, G.: 'Nonlinear adaptive generalized predictive control and applications', in CLARKE, D. (Ed.): 'Advances in model-based predictive control' (Oxford University Press, Oxford, UK, 1994), pp. 498–515
24 PARKER, R.S., DOYLE III, F.J., and PEPPAS, N.A.: 'A model-based algorithm for blood glucose control in type I diabetic patients', *IEEE Trans. Biomed. Eng.*, 1999, **46** (2), pp. 148–57
25 FLIESS, M., LEVINE, J., MARTIN, P., and ROUCHON, P.: 'Flatness and defect of nonlinear systems: introductory theory and examples', *Int. J. Control*, 1995, **61**, pp. 1327–61

26 FLIESS, M., LEVINE, J., and ROUCHON, P.: 'On differentially at nonlinear sysytems', in FLIESS, M. (Ed.): 'Nonlinear control systems design' (Pergamon Press, Bordeaux, France, 1992), pp. 408–12
27 MAHADEVAN, R., AGRAWAL, S.K., and DOYLE III, F.J.: 'Differential algebraic approach to optimization in fed-batch bioreactors'. IFAC symposium on *Advanced control of chemical processes*, Pisa, Italy, 2000
28 SALDIVAR, E., and RAY, W.: 'Mathematical modeling of emulsion copolymerization reactors: experimental validation and application to complex systems', *Ind. Eng. Chem. Res.*, 1997, **36**, pp. 1322–36
29 SALDIVAR, E., DAFNIOTIS, P., and RAY, W.H.: 'Mathematical modelling of emulsion copolymerization reactors. I. Model formulation and application to reactors operating with micellar nucleation', *Rev. Macromol. Chem. Phys.*, 1998, **C38** (2), pp. 207–325
30 SEMINO, D., and RAY, W.: 'Control of systems described by population balance equations – I. Controllabity analysis', *Chem. Eng. Sci.*, 1995, **50** (11), pp. 1805–24
31 SEMINO, D., and RAY, W.: 'Control of systems described by population balance equations – II. Emulsion polymerization with constrained control action', *Chem. Eng. Sci.*, 1995, **50** (11), pp. 1825–39
32 DIMITRATOS, J., ELICABE, G., and GEORGAKIS, C.: 'Control of emulsion polymerisation reactors', *AIChE J.*, 1994, **40** (12), pp. 993–2021
33 LIOTTA, V., GEORGAKIS, C., and EL-AASSER, M.: 'Controllability issues concerning particle size in emulsion polymerization'. Proceedings of DYCORD, 1995, pp. 299–304
34 LIOTTA, V., GEORGAKIS, C., and EL-AASSER, M.: 'Real-time estimation and control of particle size in semi-batch emulsion polymerization'. Proceedings of American Control Conference, Albuquerque, NM, 1997, pp. 1172–6
35 LIOTTA, V., GEORGAKIS, C., SUDOL, E.D., and EL-AASSER, M.S.: 'Manipulation of competitive growth for particle size control in emulsion polymerization', *Ind. Eng. Chem. Res.*, 1997, **36**, pp. 3252–63
36 LIOTTA, V., SUDOL, E.D., EL-AASSER, M.S., and GEORGAKIS, C.: 'On-line monitoring, modelling, and model validation of semibatch emulsion polymerization in an automated reactor control facility', *J. Polymer Sci.*, 1998, **36**, pp. 1553–71
37 COEN, E.M., GILBERT, R.G., MORRISON, B.R., LEUBE, H., and PEACH, S.: 'Modeling particle size distributions and secondary particle formation in emulsion polymerisation', *Polymer*, 1998, **39** (26), pp. 7099–112
38 DAFNIOTIS, P.: 'Modelling of emulsion copolymerization reactors operating below the critical micelle concentration'. PhD Thesis, The University of Wisconsin at Madison, 1996
39 ECHEVARRIA, A., LEIZA, J., DE LA CAL, J., and ASUA, J.: 'Molecular-weight distribution control in emulsion polymerization', *AIChE J.*, 1998, **44** (7), pp. 1667–79
40 GILBERT, R.G.: 'Emulsion polymerization: a mechanistic approach' (Academic Press, San Diego, 1995)
41 GILBERT, R.G.: 'Modelling rates, particle size distributions and molar mass distributions', in LOVELL, P.A., and EL-AASSER, M.S. (Eds): 'Emulsion polymerization and emulsion polymers' (John Wiley & Sons, Ltd, 1997), chap. 5, pp. 165–203

42 RAWLINGS, J., and RAY, W.: 'The modeling of batch and continuous emulsion polymerization reactors. Part II: Comparison with experimental data from continuous stirred tank reactors', *Polym. Eng. Sci.*, 1988, **28**, pp. 257–74

43 RAWLINGS, J., and RAY, W.: 'The modeling of batch and continuous emulsion polymerization reactors. Part I: Model formulation and sensitivity to parameters', *Polym. Eng. Sci.*, 1998, **28**, pp. 237–56

44 SALDIVAR, E.: 'Modeling and control of emulsion copolymerization reactors'. PhD Thesis, University of Wisconsin-Madison, 1996

45 BAINES, M.J.: 'Moving finite elements', in 'Monographs on numerical analysis' (Clarendon Press, Oxford, 1994)

46 FINLAYSON, B.A.: 'Nonlinear analysis in chemical engineering' (McGraw-Hill, New York, 1980)

47 VILLADSEN, J., and MICHELSEN, M.: 'Solution of differential equation models by polynomial approximation' (Prentice-Hall, Englewood Cliffs, NJ, 1978)

48 DOS RAMOS, J.G., and SILEBI, C.A.: 'Submicron particle size and polymerization excess surfactant analysis by capillary hydrodynamic fractionation (CHDF)', *Polymer Int.*, 1993, **30** (4), pp. 445–50

49 BANERJEE, A.B., ARKUN, Y., OGUNNAIKE, B., and PEARSON, R.: 'Estimation of nonlinear systems using linear multiple models', *AIChE J.*, 1997, **43** (5), pp. 1204–26

50 FLAUS, J.M., and BOILLEREAUX, L.: 'Moving horizon state estimation for a bioprocess modelled by a neural network', *Trans. Inst. Meas. Control*, 1997, **19** (5), pp. 263–70

51 OHTSUKA, T., and FUJII, H.A.: 'Nonlinear receding-horizon state estimation by real-time optimization technique', *J. Guid. Control Dyn.*, 1996, **19** (4), pp. 863–70

52 RAO, C., and RAWLINGS, J.: 'Nonlinear moving horizon state estimation'. International symposium on *Nonlinear model predictive control: assessment and future directions*, Ascona, Switzerland, 1998, pp. 146–63

53 RUSSO, L.P., and YOUNG, R.E.: 'Moving-horizon state estimation applied to an industrial polymerization process'. Proceedings of American Control Conference, San Diego, CA, 1999, pp. 1129–33

54 YAZ, E., and YILDIZBAYRAK, N.: 'Moving horizon control and moving window estimation schemes for discrete time-varying systems', *Int. J. Sys. Sci.*, 1987, **18** (8), pp. 1447–56

55 BEMPORAD, A., MIGNONE, D., and MORARI, M.: 'Moving horizon estimation for hybrid systems and fault detection'. Proceedings of American Control Conference, San Diego, CA, 1999, pp. 2471–5

56 MIGNONE, D., BEMPORAD, A., and MORARI, M.: 'A framework for control, fault detection, state estimation, and verification of hybrid systems'. Proceedings of American Control Conference, San Diego, CA, 1999, pp. 134–8

57 TYLER, M.L., and MORARI, M.: 'Qualitative modelling using propositional logic'. AIChE Fall National Meeting, Chicago, 1996

58 TYLER, M.L., and MORARI, M.: 'Propositional logic in control and monitoring problems', *Automatica*, 1999, **35**, pp. 565–82

59 CHEN, H., KREMLING, A., and ALLGÖWER, F.: 'Nonlinear predictive control of a benchmark CSTR'. Proceedings of European Control Conference, Rome, Italy, 1995, pp. 3247–52

60 ENGELL, S., and KLATT, K.-U.: 'Nonlinear control of a non-minimum-phase CSTR'. Proceedings of American Control Conference, San Francisco, CA, 1993, pp. 2941–5

2.9 Notes

[1]SEKI, H., and MORARI, M.: 'Optimal control of processes with a singular static gain matrix', *Automatica* (submitted, 1998)

Part II

Chapter 3

Model predictive control: output feedback and tracking of nonlinear systems

L. Magni, G. De Nicolao and R. Scattolini

Abstract

This chapter deals with nonlinear model predictive control algorithms solving the output-feedback and tracking problems for nonlinear systems. First, it is shown how to combine a generic state-feedback exponentially stabilising controller with a stable nonlinear observer to obtain a stabilising output-feedback controller. Then, three algorithms are proposed to achieve asymptotic zero-error tracking for different classes of reference signals and system representations.

3.1 Introduction

The model predictive control (MPC) algorithms that are widely and effectively adopted in industrial process control usually rely on linear plant models [1–6]. The extension of the MPC strategy to the control of plants described by nonlinear models is appealing for a number of reasons, among which one may cite the regulation around different setpoints and the transition from one setpoint to another without the need for designing (and tuning) a different controller for each linearised model. Since receding horizon (RH) optimisation is the core of MPC, the first step towards the development of usable nonlinear MPC schemes was the investigation of the state-feedback RH control problem for nonlinear systems. At present, there are several algorithms that yield a stabilising control law. Apart from algorithms that impose a terminal zero-state constraint [7, 8], and those based on the contractive mapping notion [9, 10], most methods rely on a locally stabilising controller (e.g. a linear one), an associated invariant set and a terminal penalty. The dual mode controller uses finite horizon optimisation to drive the state inside the invariant set and then switches to the local controller [11]. Alternatively, given a locally stabilising linear control law, closed-loop stability can be enforced by a proper joint choice of an ellipsoidal invariant set and a quadratic terminal penalty [12, 13]. Finally, one can also use a nonquadratic terminal penalty equal to the cost-to-go incurred closing the loop with the locally stabilising control law [14].

The second step, that is output feedback and tracking, is not as well established. Among the existing results there is the continuous-time output feedback scheme (without tracking) obtained by complementing the dual model controller with a newly designed moving horizon observer [15]. Another feature of 'standard' MPC algorithms that should be extended to the nonlinear case is the ability to track preprogrammed reference signals.

The aim of this chapter is to present some possible solutions, recently proposed in References 16–19, to the output feedback and tracking problems. First of all, it is shown that stabilising output feedback controllers can be obtained combining any stabilising state feedback nonlinear RH control law with a stable nonlinear observer, for example the classical extended Kalman filter [17]. Starting from this result and relying on some recent developments of nonlinear control theory [20], an output feedback MPC algorithm is presented for the tracking of exogenous signals (not necessarily constant) and the asymptotic rejection of disturbances generated by a properly defined exosystem [16]. Although this method solves the tracking and disturbance rejection problems for a wide class of persistent signals, its applicability is hampered by the computational load. In fact, it is necessary to compute 'reference trajectories' for the state and the input.

For this reason, another algorithm is introduced which guarantees stability and asymptotic tracking of a class of signals widely used in industrial applications, herein called 'asymptotically' constant signals. The idea is to consider the change from a working point to another one along a preprogrammed trajectory. Then, at the end of a finite horizon the reference signal is constant and the nominal equilibrium of the system corresponding to the final constant reference can be easily computed. The method is based on the solution of a finite-horizon optimisation problem where the future errors and control deviations are minimised with respect to the future control actions (possibly subject to constraints) until the reference signal becomes constant. At that point, a terminal penalty and a terminal stability region are imposed following one of the stabilising algorithms available for the state-feedback regulation problem. In order to reduce the computational load, one can optimise only the first part of the future control actions while the subsequent ones are computed by means of a stabilising linear dynamic regulator. Notably, as a consequence of the RH approach, this regulator is never applied in practice, but is only used to evaluate the performance index. The main features of the method are the following: (a) it does not call for the computation of the input and state trajectories when the reference changes; (b) it allows for the use of preprogrammed references; (c) it takes advantage of a locally stabilising linear regulator, which can be the result of a linear design procedure applied on the linearised plant.

Finally, for the same class of reference signals an algorithm is proposed for systems described by an input/output nonlinear autoregressive exogenous (NARX) model [18]. The main advantage of this approach is that the use of an observer is not necessary because the control law is synthesised directly on the input/output model.

3.2 Preliminaries and state-feedback control

In this section, we briefly introduce a fairly general formulation of RH state feedback control for nonlinear systems. The symbol $\|\cdot\|$ denotes the Euclidean norm in $\Re^n$ (where the dimension n follows from the context); B_r denotes the closed ball of radius r, i.e. $B_r = \{x \in \Re^n : \|x\| \leq r\}$. A function $W : \Re \to \Re_+$ is said to belong to class K_∞ if (a) it is continuous, (b) $W(s) = 0 \Leftrightarrow s = 0$, (c) it is nondecreasing and (d) $W(s) \to \infty$ when $s \to \infty$.

The nonlinear discrete-time dynamic system to be controlled is

$$x(k+1) = f(x(k), u(k)), \quad x(t) = \bar{x} \quad k \geq t \tag{1}$$

$$y(k) = h(x(k)) \tag{2}$$

where $x \in \Re^n$ is the state, $y \in \Re^m$ is the output and $u \in \Re^m$ is the input. The functions $f(\cdot, \cdot)$ and $h(\cdot)$ are C^2 functions of their arguments and $f(0, 0) = 0$, $h(0) = 0$.

We search for a control law $u = \kappa(x)$ which regulates the state of (1) to the origin, subject to the input and state constraints

$$x(k) \in X, \quad u(k) \in U, \quad k \geq t \tag{3}$$

where X and U are closed subsets of $\Re^n$ and $\Re^m$, respectively, both containing the origin as an interior point. With reference to system (1) subject to the control law $u = \kappa(x)$, the origin is said to be an exponentially stable equilibrium if there exist finite constants $r, a, b > 0$ such that, $\forall \bar{x} \in B_r$,

$$\|x(k)\| \leq a\|\bar{x}\|e^{-b(k-t)}, \quad \forall k \geq t \tag{4}$$

Given a general locally stabilising control law $u = \kappa(x)$, e.g. a linear control law synthesised on the linearised system, the following finite horizon optimisation problem is introduced.

Finite-horizon optimal control problem I (FHOCP I)
Given the positive integers N_c, N_p, $N_c \leq N_p$ and a control law $u = \kappa(x)$ that stabilises the origin, minimise with respect to $u_{t,t+N_c-1} := [u(t)u(t+1)\ldots$

$u(t+N_c-1)]$, $N_c \geq 1$, the cost function

$$J(\bar{x}, u_{t,t+N_c-1}, N_c, N_p) = \sum_{k=t}^{t+N_p-1} \psi(x(k), u(k)) + V_f(x(t+N_p)) \tag{5}$$

$$\psi(x, u) = x^T Q x + uRu, \quad Q > 0, \quad R > 0$$

subject to (1), (3), $u = \kappa(x)$, $k \in [t+N_c, t+N_p-1]$, and with terminal state constraint

$$x(t+N_p) \in X_f \tag{6}$$

In (5), V_f is a suitable terminal state penalty, N_p is the prediction horizon and N_c is the control horizon. In (6), $X_f \subseteq \Re^n$ is a suitable set including the origin as an interior point.

Given an initial state $\bar{x} \in X$ the sequence $u_{t,t+N_c-1}$ is termed admissible if, when applied to system (1) with $u = \kappa(x)$, $k \in [t+N_c, t+N_p-1]$,

$$x(k) \in X, \quad u(k) \in U, \quad t \leq k < t+N_p$$

$$x(t+N_p) \in X_f$$

Associated with FHOCP I, the following control strategy is formulated.

Nonlinear Receding Horizon (NRH) control law
At every time instant t let $\bar{x} = x(t)$ and find an admissible control sequence $u^o_{t,t+N_c-1}$ solving FHOCP I. Then apply the control law $u = u^0(\bar{x})$, where $u = u^0(\bar{x})$ is the first element of $u^o_{t,t+N_c-1}$.

The NRH control law is defined by the function $u = \kappa^{NRH}(x)$, where $\kappa^{NRH}(x) = u^o(x)$. Let $X^I_0(N_c, N_p)$ be the set of states $\bar{x}$ such that there exists an admissible control sequence $u^o_{t,t+N_c-1}$ solving FHOCP I.

Assumption A1: Given Q, R, N_c, N_p, assume that the terminal penalty $V_f(\cdot)$ and the set X_f are such that the closed-loop system

$$x(k+1) = f(x(k), \kappa^{NRH}(x(k)))$$

admits the origin as an exponentially stable equilibrium with $X^I_0(N_c, N_p)$ as the domain of attraction with constrainments fulfillment.

Remark 1: *There are several ways to design $V_f(\cdot)$ and X_f so as to comply with Assumption A1. Historically, the first RH method with guaranteed stability was based on the choice $N_c = N_p$, $V_f(x) \equiv 0$ and $X_f = \{0\}$; see Reference 21 for linear systems and References 7, 8 and 22–24 for nonlinear systems. In this way, the FH minimisation problem becomes equivalent to the minimisation of (5) subject to (1), (3) and the terminal constraint $x(t+N_p) = 0$.*

Given a local control law $u = \kappa(x)$ that stabilises the origin, a second possible choice for the terminal region X_f is a positively invariant set which is a domain of attraction of the origin for the system

$$x(k+1) = f(x(k), \kappa(x(k))) \tag{7}$$

and such that $\bar{x} \in X_f$ implies that $x(k) \in X$, $\kappa(x(k)) \in U$, $k \geq t$. In other words, X_f must be an output admissible set for the closed-loop system (7) [25]. In particular, a possible $\kappa(x)$ is the optimal linear quadratic gain K^{LQ} for the linearised system, namely $\kappa(x) = K^{LQ}x$. Then, an output admissible set for the closed loop (7) can be computed as a suitable level set of a Lyapunov function for the linearised system; see References 11, 13 and 26.

As for $V_f(\cdot)$, in References 12 and 13 it has been shown that there exist choices of $P_f = P_f^T > 0$ such that $V_f(x) = x^T P x$ guarantees closed-loop stability of the associated RH scheme.

Alternatively, one can use a (generally nonquadratic) terminal penalty $V_f(\cdot)$ defined as

$$V_f(\bar{x}) = \sum_{k=t}^{t+M} \{x^T(k)Qx(k) + x^T(k)K^{LQT}RK^{LQ}x(k)\} \tag{8}$$

subject to (1) with $u(k) = K^{LQ}x(k)$. In other words, $V_f(\bar{x})$ in (8) is the cost that is incurred by applying the LQ control law for $k \in [t, t+M]$. This RH scheme has been recently proposed, with $M = \infty$ [4, 14, 27, 28] and has some analogies with ideas originally developed in the context of LQ control with constraints [29–32]. Finally, Reference 33 introduced a computable algorithm with a finite horizon M.

For a critical survey of different state-feedback algorithms see, for example, References 34–36.

Remark 2: *The use of a very short control horizon N_c and a relatively long prediction horizon N_p, though usually neglected in the nonlinear predictive literature, is a very effective feature of 'classical' predictive control. In fact, the computational load is strictly related to the control horizon N_c, whereas stability and performance can be governed using the prediction horizon N_p. When a finite prediction and control horizon are used, the actual closed-loop input- and state-trajectories will in general differ from the predicted open-loop trajectories even if no model plant mismatch and no disturbances are present. Assume that our goal is to compute a feedback minimising a performance objective of the type (5) but extended over the infinite horizon. It is by no means clear that repeatedly minimising the finite horizon objective (5) in a receding horizon fashion leads to an optimal solution for the infinite horizon problem. Along this direction, the choice of a long prediction horizon can reduce the difference between actual and predicted trajectories forcing the RH control law to resemble more closely the infinite-horizon optimal one.*

3.3 Output feedback

In practice the whole state vector is usually not available for feedback. Therefore, the problem of output feedback control is essential from an application viewpoint. In this section, it is shown how to produce a stabilising dynamic output feedback regulator by combining a stabilising state feedback control law with an asymptotic state observer. To this purpose consider the following state detector for system (1)–(2):

$$\hat{x}(k+1) = g(\hat{x}(k), y(k), u(k)), \quad \hat{x}(t) = \bar{\hat{x}}, \quad k \geq t \tag{9}$$

Definition 1 [37]: *The system (1)–(2) is weakly detectable if there exists a function* $g : \Re^n \times \Re^m \times \Re^m \to \Re^n$ *and a function* $V_E \in C^1$ *such that the following properties hold:*

1. $g(\hat{x}, y, u)$ *is continuous and* $g(0, 0, 0) = 0$.
2. *There exist functions* a_E, b_E, c_E *of class* K_∞ *and a positive real number E such that*

$$a_E(\|x - \hat{x}\|) \leq V_E(x, \hat{x}) \leq b_E(\|x - \hat{x}\|), \quad \forall (x, \hat{x}) \in B_E \times B_E \tag{10}$$

$$\Delta V_E(x, \hat{x}) \leq -c_E(\|x - \hat{x}\|), \quad \forall (x, \hat{x}, u) \in B_E \times B_E \times B_E \tag{11}$$

where $\Delta V_E(x, \hat{x}) := V_E(f(x, u), g(\hat{x}, y(x), u)) - V_E(x, \hat{x})$.

If (9) is a 'weak detector' for system (1)–(2), then, provided that $u(k) \in B_E$ and the solution trajectories $x(k)$ and $\hat{x}(k)$ do not leave B_E, $[x(k) - \hat{x}(k)] \to 0$ as $k \to \infty$.

Definition 2 [38]: *The system (1)–(2) is weakly exponentially detectable if one can find a function* $g : \Re^n \times \Re^m \times \Re^m \to \Re^n$ *and a function* V_E *satisfying (10)–(11), with* $a_E(\sigma)$, $b_E(\sigma)$, $c_E(\sigma)$ *of the form* $a_E(\sigma) = a_E\sigma^n$, $b_E(\sigma) = b_E\sigma^n$, $c_E(\sigma) = c_E\sigma^n$ *for some positive constants* a_E, b_E, c_E *and* η.

By analogy with the comment after Definition 1, it is easy to verify that, if (9) is a weak exponential detector for system (1)–(2), then $[x(k) - \hat{x}(k)] \to 0$ exponentially as $k \to \infty$.

The design of stable nonlinear observers has been the object of recent studies in the literature. For instance, state observers based on the minimisation of a moving horizon cost function have been presented in References 15 and 39–41 for continuous and discrete time systems, respectively. Furthermore, the convergence properties of the popular extended Kalman filter (EKF) have been analysed in Reference 42. Since the following results on output feedback control hold true for any weak detector of system (1)–(2), we shall not concentrate on a specific choice of the observer. Then, the following assumption is introduced.

Assumption A2: The system (1)–(2) is weakly detectable and a weak detector is used.

The dynamic regulator composed by a state observer and a state feedback control law $u = \kappa(x)$ (it is hereafter assumed that $\kappa(0) = 0$) is then described by

$$x(k+1) = g(\hat{x}(k), h(x(k)), u(k)), \quad \hat{x}(t) = \bar{\hat{x}} \tag{12}$$

$$u(k) = \kappa(\hat{x}(k)) \tag{13}$$

where $g(\hat{x}, y, u)$ is a weak detector of system (1)–(2) and $\bar{\hat{x}}$ is an initial guess of the state $x(t)$.

Concerning the system (1), (2), (12), (13) the following assumptions are in order.

Assumption A3: There exist positive real numbers L_κ, L_g, ρ and E such that

$$\frac{\|\kappa(x_1) - \kappa(x_2)\|}{x_1 - x_2} \leq L_\kappa, \quad \forall x_1, x_2 \in B_\rho \tag{14}$$

$$\max\left\{ \frac{\|g(x_1, y_1, u_1) - g(x_2, y_2, u_2)\|}{x_1 - x_2}, \frac{\|g(x_1, y_1, u_1) - g(x_2, y_2, u_2)\|}{y_1 - y_2}, \frac{\|g(x_1, y_1, u_1) - g(x_2, y_2, u_2)\|}{u_1 - u_2} \right\} \leq L_g \quad \forall x_1, x_2, u_1, u_2, y_1, y_2 \in B_E \tag{15}$$

The main stability results of this section extend to the discrete-time case results in Reference 38.

Thereom 1 [17]: Assume that Assumptions A2 and A3 hold and the state feedback control law $\kappa(x)$ exponentially stabilises the origin of (1). Then, the origin in $\Re^n \times \Re^n$ is an asymptotically stable equilibrium point of the closed-loop system (1)–(2), (12)–(13).

Thereom 2 [17]: Suppose that the assumptions of Theorem 1 are satisfied and that system (1)–(2) is weakly exponentially detectable, and an exponential detector is used. Then, the origin in $\Re^n \times \Re^n$ is an exponentially stable equilibrium point of the closed-loop system (1)–(2), (12)–(13).

Remark 3: *For conciseness, the characterisation of the region of attraction has not been reported in the statements of Theorems 1 and 2. However, a finite region of attraction, whose extent depends on the scalars ρ and E that characterise the regions of attraction of the controller and the observer, is explicitly defined in Reference 17. For this reason the results of Theorems 1 and 2 are not merely local. The Lipschitz assumptions (14), (15), though strong, are not easily relaxable; see, for example, Reference 37, where similar assumptions are introduced in the continuous time case. Unfortunately, it is not easy to find conditions guaranteeing that the RH scheme of the previous section yields a Lipschitz continuous control law so that (14) must be assumed; see also the discussion in Reference 24 (Remark 9).*

3.4 Tracking and disturbance rejection for signals generated by an exosystem

This section deals with the servomechanism problem, that is, designing a regulator for system (1)–(2) that guarantees asymptotic tracking and disturbance rejection for a class of persistent reference signals and disturbances. Consider the following nonlinear, discrete-time dynamic system:

$$\begin{aligned} x(k+1) &= f(x(k), u(k), w(k)) \\ e(k) &= h_e(x(k), w(k)) \end{aligned} \tag{16}$$

where $f(x, u, w)$, $h_e(x, w)$ are smooth (i.e. C^∞) functions with $f(0, 0, 0) = 0$, $h_e(0, 0) = 0$ and $w \in \Re^r$ is a set of exogenous input variables which includes disturbances $d(\cdot)$ (to be rejected) and/or references $y^0(\cdot)$ (to be tracked). In both cases, the matter is to impose that the tracking error $e(k) = y^0(k) - y(k)$ (which, with a proper choice of h_e can be written as $e(k) = h_e(x(k), w(k))$) decays to zero asymptotically, for every output and every disturbance taken from prespecified families of functions. In particular, it is assumed that $w(\cdot)$ is the solution of a (possibly nonlinear) homogeneous dynamic exosystem

$$w(k+1) = s(w(k)) \tag{17}$$

where $s(w)$ is a C^∞ function with $s(0) = 0$ and the initial condition $w(0)$ belongs to some neighbourhood W of the origin in $\Re^r$. Concerning the exosystem (17), the following assumption is made in this chapter.

Assumption A4: The exosystem is neutrally stable.

For the definition of neutral stability, the reader is referred to Reference 20 (p. 388). Note that A4 implies that all the eigenvalues of the matrix

$$S = \frac{\partial s}{\partial w}(0)$$

are on the unit circle.

The following output-feedback predictive control law solving the servomechanism problem relies on some results due to Isidori and Byrnes [43]. According to Reference 20, we are now in the position to define the following problem.

Output tracking problem (OTP)
Given a nonlinear system of the form (16) and a neutrally stable exosystem (17) find a dynamic regulator

$$\begin{aligned} z(k+1) &= \eta(z(k), e(k)) \\ u(k) &= \theta(z(k)) \end{aligned}$$

with z defined in a neighbourhood Z of the origin in $\Re^\nu$ such that:

1. The linearisation of

$$\begin{aligned} x(k+1) &= f(x(k), \theta(z(k)), 0) \\ z(k+1) &= \eta(z(k), h_e(x(k), 0)) \end{aligned}$$

at the equilibrium $(x, z) = (0, 0)$ is asymptotically stable.
2. There exists a neighbourhood $V \subset X \times Z \times W$ of $(0, 0, 0)$ such that, for any initial condition $(x(0), z(0), w(0)) \in V$, the solution of

$$\begin{aligned} x(k+1) &= f(x(k), \theta(z(k)), w(k)) \\ z(k+1) &= \eta(z(k), h_e(x(k), w(k))) \\ w(k+1) &= s(w(k)) \end{aligned}$$

satisfies

$$\lim_{k \to \infty} h_e(x(k), w(k)) = 0.$$

Let

$$A = \frac{\partial f}{\partial x}(0, 0, 0), \quad B = \frac{\partial f}{\partial u}(0, 0, 0), \quad C = \frac{\partial h_e}{\partial x}(0, 0)$$

The following theorem is the extension to the discrete time case of Theorem 8.4.4 in Reference 20.

Thereom 3: The output tracking problem is solvable if and only if there exist mappings $x = \pi(w)$ and $u = c(w)$, with $\pi(0) = 0$ and $c(0) = 0$, both defined in a neighbourhood $W^0 \subseteq W$ of the origin, satisfying the conditions:

$$\begin{aligned} \pi(s(w)) &= f(\pi(w), c(w), w) \\ 0 &= h_e(\pi(w), w) \end{aligned} \tag{18}$$

for all $w \in W^0$, and such that the autonomous system $\{W^0, s, c\}$ through the map $\xi = \tau(w)$ is immersed (see Reference 20) into a system:

$$\begin{aligned} \xi(k+1) &= \varphi(\xi(k)) \\ ; u(k) &= \gamma(\xi(k)) \end{aligned}$$

defined in a neighbourhood Ξ^0 of the origin in $\Re^\nu$, in which $\varphi(0) = 0$ and $\gamma(0) = 0$, and the two matrices

$$\Phi = \frac{\partial \varphi}{\partial \xi}(0), \quad \Gamma = \frac{\partial \gamma}{\partial \xi}(0)$$

are such that the pair

$$\begin{bmatrix} A & 0 \\ TC & \Phi \end{bmatrix}, \begin{bmatrix} B \\ 0 \end{bmatrix} \tag{19}$$

is stabilisable for some choice of the matrix T, and the pair

$$[C \quad 0], \begin{bmatrix} A & B\Gamma \\ 0 & \Phi \end{bmatrix} \tag{20}$$

is detectable.

It is worth pointing out that Theorem 3 is only a local result.

Note that, as observed in Reference 20, (19) and (20) imply that the triplet

$$(\bar{A}, \bar{B}, \bar{C}) = \left(\begin{bmatrix} A & B\Gamma \\ TC & \Phi \end{bmatrix}, \begin{bmatrix} B \\ 0 \end{bmatrix}, [C \quad 0] \right)$$

is stabilisable and detectable. Conditions guaranteeing the stabilisability of the pair (19) and the detectability of the pair (20) are given in Reference 20 in terms of the original data (A, B, C) and the pair (Φ, Γ).

The output tracking problem is now solved using a nonlinear receding horizon control law. To this purpose, first consider the system composed by (16)–(17) and an internal model of the exosystem:

$$\begin{aligned} x(k+1) &= f(x(k), u(k), w(k)) \\ z_1(k+1) &= \varphi(z_1(k)) + Te(k) \\ w(k+1) &= s(w(k)) \\ e(k) &= h_e(x(k), w(k)) \\ u(k) &= \gamma(z_1(k)) + v(k) \end{aligned} \tag{21}$$

where T is an appropriate matrix and $\varphi(\cdot) : \Re^\nu \to \Re^\nu$ and $\gamma(\cdot) : \Re^\nu \to \Re^m$ are those of Theorem 3. The signal $v(k)$ is the part of the input that will be used to solve the tracking problem. Let w_M and w_U be the measurable and the unmeasurable components of w so that

$$w(k+1) = \begin{bmatrix} w_U(k+1) \\ w_M(k+1) \end{bmatrix} = \begin{bmatrix} s_U(w_U(k), w_M(k)) \\ s_M(w_M(k)) \end{bmatrix}$$

where s_M and s_U are suitable functions. Define the states of system (21) to be detected as

$$x_D = \begin{bmatrix} x^T & z_1^T & w_U^T \end{bmatrix}^T$$

Then, in view of (21), one can write

$$\begin{aligned} x_D(k+1) &= f_D(x_D(k), v(k), w_M(k)), \quad x_D(t) = \bar{x}_D, \quad k \geq t \\ e(k) &= h_D(x_D(k), w_M(k)) \end{aligned} \tag{22}$$

for suitable functions f_D, h_D, and where $\bar{x}_D$ is defined in a obvious way.

Now assume that system (22) is weakly detectable and set up a weak detector for the state x_D:

$$\hat{x}_D(k+1) = g\left(\hat{x}_D(k), e(k), \begin{bmatrix} v(k) \\ w_M(k) \end{bmatrix}\right)$$

Let also

$$z := \begin{bmatrix} x_D' & w_M' & z_1^{c'} \end{bmatrix}', \quad \hat{z} := \begin{bmatrix} \hat{x}_D' & w_M' & z_1' \end{bmatrix}$$

where $z_1^c \in \Re^{\nu}$ is used to emulate the internal model with the known initial condition z_1 in order to satisfy the constraints on u. Obviously, $z \in \Re^{n+2\nu+r}$ and $\hat{z} \in \Re^{n+2\nu+r}$. In the following, Z will be a closed subset in $\Re^{n+2\nu+r}$ containing the origin as an interior point. Letting $\zeta(i) = [I_{n+\nu} \; 0_{\nu+r}]z(i)$, where I_n and O_n are row vectors with all the n elements equal to one and zero respectively, in order to synthesize the feedback *RH* control law $\nu = \kappa^{NRH}(\hat{z}(k))$, consider the following problem.

Finite horizon optimal control problem II (FHOCP II):
Given the positive integers N_c, N_p, $N_c \leq N_p$ and a locally stabilising control law $\nu = \kappa(z)$, minimise with respect to $\nu_{t,t+N_c-1}$, $N_c \geq 1$, the cost function

$$J(\hat{z}(t), \nu_{t,t+N_c-1}, N_c, N_p) = \sum_{k=t}^{t+N_p-1} \psi\left(\zeta(k) - \begin{bmatrix} \pi(w(k)) \\ \tau(w(k)) \end{bmatrix}, \nu(k)\right) + V_f(z(t+N_p)) \tag{23}$$

subject to eqn (21) and

$$\begin{aligned} z_1^c(k+1) &= \varphi(z_1^c(k)) + Te(k) \\ u^c(k) &= \gamma(z_1^c(k)) + \nu(k) \end{aligned}$$

with initial state $\hat{z}(t)$, to the constraints $z(i) \in Z$, $u^c(k) \in U$, $t \leq k < t + N_p - 1$, $\nu(k) = \kappa(z(k))$, $t + N_c \leq k < t + N_p - 1$, and to the terminal state constraint

$$z(t+N_p) \in X_f$$

where V_f and $X_f \in \Re^{n+2\nu+r}$ are a suitable set.

Based on *FHOCP II* the *NRH* control law $v = \kappa^{NRH}(\hat{z})$ is obtained in the usual way.

For the overall controller given by

$$\begin{aligned} \hat{x}_D(k+1) &= g\Big(\hat{x}_D(k), e(k), \left[\, \kappa^{NRH}(\hat{z}(k))' \;\; w_M(k)' \,\right]'\Big) \\ z_1(k+1) &= \varphi(z_1(k)) + Te(k) \\ u(k) &= \gamma(z_1(k)) + \kappa^{NRH}(\hat{z}(k)) \end{aligned} \tag{24}$$

the main result of this section can now be stated.

Thereom 4 [16]: Consider system (16) and suppose that Assumptions A1 and A3–A4 hold, system (22) is weakly exponentially detectable and the output tracking problem is solvable. Then, the nonlinear receding-horizon controller (24) solves the output tracking problem.

Remark 4: *At first sight, estimating the unmeasurable disturbance state w_U could seem redundant, since disturbance rejection and tracking are guaranteed by the structure of the internal model. However, the predictive performance index (23) calls for the knowledge of the future values of such states. On the other hand, the use of future disturbances and references is a standard feature of predictive control algorithms. Analogously, estimating z_1 is necessary to guarantee that the value used in the control law is equal to the one required in nominal conditions notwithstanding the presence of modelling errors or disturbances, i.e. the robustness of the regulation action.*

3.5 Tracking 'asymptotically' constant references

In the previous section, an output feedback MPC algorithm has been presented for the tracking of exogenous signals (not necessarily constant) and for the asymptotic rejection of disturbances generated by a properly defined exosystem. In this section, the tracking problem for preprogrammed set-points, which are the most used in industrial applications, is considered. In particular we will assume that the reference signal y^0 is 'asymptotically' constant, i.e. it is constant beyond a prescribed future horizon N_r:

$$y^0(k) = \bar{y}^0, \quad \forall k \geq t + N_r \tag{25}$$

The main goal of this section is to derive a nonlinear stabilising predictive control algorithm that requires only the desired (output) trajectory. In fact, state trajectory generation (for a nonlinear system) is an old but still challenging problem and the presence of constraints imposes additional difficulties in this respect.

Two solutions are proposed. The first is based on a state-space model of the system under control. If the state of the system is not available one can use a suitable observer as shown in Section 3.3. The second is based on an input/output representation of the system under control.

3.5.1 State-space models

Consider a nonlinear dynamic system:

$$\begin{aligned} \xi(k+1) &= \tilde{f}(\xi(k), u(k)) \\ y(k) &= \tilde{h}(\xi(k)) \end{aligned} \tag{26}$$

where k is the discrete time index, $\xi \in \Re^n$ represents the system state, $u \in \Re^m$ is the input vector, $y \in \Re^m$ is the output vector, and $\tilde{f}(\cdot, \cdot)$ and $\tilde{h}(\cdot)$ are C^2 functions of their arguments. Assume now that for a given vector $y^0 \in \Re^m$ there exists an equilibrium $(\bar{\xi}(\bar{y}^0), \bar{u}(\bar{y}^0))$ such that:

$$\begin{aligned} \bar{\xi}(y^0) &= \tilde{f}(\bar{\xi}(\bar{y}^0), \bar{u}(\bar{y}^0)) \\ \bar{y}^0 &= \tilde{h}(\bar{\xi}(\bar{y}^0)) \end{aligned}$$

and let

$$\tilde{A}(\bar{y}^0) = \left.\frac{\partial \tilde{f}}{\partial \xi}\right|_{\bar{\xi}(\bar{y}^0), \bar{u}(\bar{y}^0)}, \quad \tilde{B}(\bar{y}^0) = \left.\frac{\partial \tilde{f}}{\partial u}\right|_{\bar{\xi}(\bar{y}^0), \bar{u}(\bar{y}^0)}, \quad \tilde{C}(\bar{y}^0) = \left.\frac{\partial \tilde{h}}{\partial u}\right|_{\bar{\xi}(\bar{y}^0), \bar{u}(\bar{y}^0)}$$

In order to give conditions ensuring the solvability of the problem, the following assumption is introduced.

Assumption A5: The linear system $(\tilde{A}(\bar{y}^0), \tilde{B}(\bar{y}^0), \tilde{C}(\bar{y}^0))$ is reachable, observable and does not possess transmission zeros equal to one.

As usual in both linear and nonlinear control theory, asymptotic tracking and disturbance rejection problems can be solved provided that the regulator includes an internal model of the exosystem generating the exogenous signals. In this case, since 'asymptotically' constant exogenous signals are considered, an integrator is plugged in front of the system. Then, the control variable $u(k)$ is given by:

$$\begin{aligned} z(k+1) &= z(k) + \Delta u(k) \\ u(k) &= z(k) \end{aligned} \tag{27}$$

and the overall system composed by (26) and (27) is described by:

$$x(k+1) = f(x(k), \Delta u(k)), \quad x(t) = \hat{x}, \quad k \geq t \tag{28}$$

$$y(k) = h(x(k)) \tag{29}$$

$$e(k) = y^0 - y(k) \tag{30}$$

where $f(\cdot, \cdot)$ is defined in an obvious way and $x = [\xi^T z^T]^T$. Moreover, define

$$\bar{x}(\bar{y}^0) = \begin{bmatrix} \bar{\xi}(\bar{y}^0) \\ \bar{u}(\bar{y}^0) \end{bmatrix}$$

For the augmented system (28)–(29) the following well-known result holds.

Lemma 5: Consider $\bar{y}^0 \in \Re^m$ and the associated equilibrium $(\bar{\xi}(\bar{y}^0), \bar{u}(\bar{y}^0))$ of (26) with $\tilde{h}(\bar{\xi}(\bar{y}^0))$. Then, under Assumption A5, for system (28)–(30) there exists a linear dynamic regulator (LDR):

$$\begin{aligned} x_r(k+1) &= A_r x_r(k) + B_r e(k), \quad x_r(t) = \hat{x}_r, \quad k \geq t \\ \Delta u(k) &= C_r x_r(k) + D_r e(k) \end{aligned} \tag{31}$$

with $x_r \in \Re^{n_r}$ such that $\bar{x}_{cl}(\bar{y}^0) = [\bar{x}^T(\bar{y}^0) \quad 0]^T$ is a stable equilibrium point of the closed-loop system (28)–(31).

Assumption A6: With reference to system (28)–(29), with $\Delta u(k) = 0$, if $y(k) = \bar{y}^0$, $\forall k \geq t$, then $x(t) = \bar{x}(\bar{y}^0)$.

Remark 5: *Under Assumption A5, from standard continuity arguments it follows that there exists a finite neighbourhood of the equilibrium point $\bar{x}(\bar{y}^0)$ where Assumption A6 is satisfied. In fact, A6 is implied by the observability of the linearisation of (28)–(29) around $(\bar{x}(\bar{y}^0), 0)$.*

In order to synthesise the RH control law consider the following problem.

Finite horizon optimal control problem III (FHOCP III)
Given the positive integers N_c, N_p, N_r, $N_c \leq N_p$, $N_r \leq N_p$, letting $\Delta u_{t,t+N_c-1} := [\Delta u(t) \quad \Delta u(t+1) \ldots \Delta u(t+N_c-1)]$, minimise with respect to $(\Delta u_{t,t+N_c-1}, x_r(t+N_c))$ the cost functional

$$\begin{aligned} J(\hat{x}, \Delta u_{t,t+N_c-1}, x_r(t+N_c), y^0) = &\sum_{k=t}^{t+N_c-1} \psi_o(e(k), \Delta u(k)) + x_r^T(t+N_c) R_{x_r} x_r(t+N_c) \\ &+ \sum_{k=t+N_c}^{t+N_p-1} \psi_c(e(k), x_r(k)) \\ &+ V_f(x(t+N_p), x_r(t+N_p)) \end{aligned} \tag{32}$$

subject to (28)–(30), with $\Delta u(k)$ given by (31) for $k \in [t+N_c, t+N_p-1]$ and the terminal state constraints

$$\begin{bmatrix} x(t+N_p) \\ x_r(t+N_p) \end{bmatrix} \in X_f(\bar{y}^0) \tag{33}$$

In (32), ψ_o and ψ_c have the form:

$$\begin{aligned}\psi_o(e(k),\Delta u(k)) &= e^T(k)Q_e e(k) + \Delta u^T(k)R_u\Delta u(k)\\ \psi_c(e(k),x_r(k)) &= e^T(k)\bar{Q}_e e(k) + \Delta u^T(k)\bar{Q}_{x_r}\Delta u(k)\\ \bar{Q}_e &= Q_e + D_r^T R_u D_r + B_r^T R_{x_r} B_r\\ \bar{Q}_{x_r} &= Q_{x_r} + C_r^T R_u C_r + A_r^T R_{x_r} A_r\end{aligned}$$

with Q_e, R_u, Q_{x_r} and R_{x_r} positive definite matrices. V_f and X_f are chosen following one of the algorithms cited in Remark 1 provided that the origin of the state-space has been replaced by the equilibrium $\bar{x}_{c1}(\bar{y}^0)$.

In order to derive the main stability result, temporarily assume that the plant state x is known. Let $x^{RH}(k)$ be the state movement of the closed-loop system

$$x^{RH}(k+1) = f(x^{RH}(k), \kappa^{RH}(x^{RH}(k))) \tag{34}$$

where $\Delta u = \kappa^{RH}(x)$ is the RH control law based on the FHOCP III with cost function (32). Denoting by $X_0^{III}(N_c, N_p, y^0)$ the set of states $\hat{x}$ for which an admissible solution to the FHOCP III exists, the following result guarantees closed-loop stability and asymptotic zero-error regulation.

Thereom 6 [19]: Consider a reference signal $y^0(\cdot)$ satisfying (25) and suppose that Assumptions A1 and A5–A6 hold. Then if $X_0^{III}(N_c, N_p, y^0)$ is nonempty, the closed-loop system (34) admits the state $\bar{x}(\bar{y}^0)$ as an asymptotically stable equilibrium point with output admissible set $X_0^{III}(N_c, N_p, y^0)$.

If the plant state is not accessible, the stabilising control law must be combined with an asymptotic state observer, and, as shown in Section 3.3, this produces a stabilising output feedback regulator.

3.5.2 Nonlinear ARX models

In this section the tracking problem for signals of type (25) is solved for input/output systems described by NARX (nonlinear autoregressive exogenous) models. Specifically, starting from a discrete-time state-space description of the system, first an integral action is inserted in front of the system in order to achieve robust zero-error regulation. Following the guidelines given in Levin and Narendra [44], it is then shown that, around the given equilibrium, the ensemble plant-integrator can always be locally described by a NARX model. This is a fundamental step in the algorithm definition, because in many practical cases the plant model is determined through identification experiments, and well established methodologies for the estimation of NARX systems are by now available; see, for example, References 45 and 46. Furthermore, a theoretical justification for the use of NARX models has also been given in References 47 and 48.

The main stability result is derived under the mild assumption that the linearisation of the original system under control is reachable and observable and does not possess transmission zeros at point one in the z plane. One of the advantages of this approach is that the use of an observer is not necessary. The following lemma shows that the ensemble plant-integrator (28)–(29) admits a local NARX representation.

Lemma 7 [18]: Assume that the system $(\tilde{A}, \tilde{B}, \tilde{C})$ obtained by linearising (26) around the equilibrium point $(\bar{\xi}(\bar{y}^0), \bar{u}(\bar{y}^0))$ such that $\bar{y}^0 = \tilde{h}(\bar{\xi}(\bar{y}^0))$ satisfies A5. Then, the system (28)–(29) can be locally realised by an input-output model of the form

$$y(k+1) = \gamma(y(k), \ldots, y(k-n), \Delta u(k-1), \ldots, \Delta u(k-n) \tag{35}$$

and such that

$$\bar{y}^0 = \gamma(\bar{y}^0, \ldots, \bar{y}^0, 0, \ldots, 0) \tag{36}$$

In view of Lemma 7, the system under control admits the NARX representation (35), which in turn, letting $e(k) = y(k) - y^0$, can be given the following (possibly nonminimal) state-space realisation:

$$x(k+1) = f(x(k), \Delta u(k), y^0) \tag{37}$$

$$e(k) = Cx(k) \tag{38}$$

with

$$x(k) = \left[e^T(k)e^T(k+1)\ldots e^T(k-n) \quad \Delta u^T(k-1)\ldots\Delta u^T(k-n)\right]^T \in \Re^{(2n+1)m}$$

where $f(\cdot, \cdot, \cdot)$is defined in an obvious way, while $C = [I\ 0\ 0] \in \Re^{m\times(2n+1)m}$. Obviously, the state x is accessible at any k and $0 = f(0, 0, \bar{y}^0)$. Letting

$$A_{\bar{y}^0} = \left.\frac{\partial f}{\partial x}\right|_{0,0,\bar{y}^0}, \quad B_{\bar{y}^0} = \left.\frac{\partial f}{\partial u}\right|_{0,0,\bar{y}^0} \tag{39}$$

in de Nicolao *et al.* [18] it is shown that it is possible to regulate system (37)–(38) locally using the linear control law $\Delta u = Kx$, where K stabilises the linearised system and can be designed using standard methods, e.g. pole-placement, linear quadratic (LQ) control or predictive control.

Consider the following finite-horizon problem.

Finite-horizon optimal control problem IV (FHOCP IV)

Given the positive integers N_c, N_p, N_r, with $N_c \leq N_p$ and $N_r \leq N_p$, minimise, with respect to the predicted control moves $\Delta u_{t,t+N_c-1} := [\Delta u(t)\Delta u(t+1)\ldots$

$\Delta u(t+N_c-1)]$, the cost functional

$$J(\hat{x}, \Delta u_{t,t+N_c-1}, N_c, N_p) = \sum_{k=t}^{t+N_c-1} \psi(e(k), \Delta u(k)) + V_f(x(t+N_p)) \tag{40}$$

subject to (37)–(38), $\Delta u(k) = Kx(k)$, $k \in [t+N_c, t+N_p-1]$, with $x(t) = \hat{x}$ and the terminal constraints

$$x(t+N_p) \in X_f(\bar{y}^0)$$

where V_f and X_f are chosen following one of the algorithms cited in Remark 1.

As usual the predictive control law is obtained following the RH paradigm and is defined by the function $\Delta u = \kappa^{RH}(x)$. Accordingly, $x^{RH}(k)$ is the state movement of the closed-loop system

$$x^{RH}(k+1) = f(x^{RH}(k), \kappa^{RH}(x^{RH}(k)), y^0) \tag{41}$$

Denoting by $X_0^{IV}(N_c, N_p, y^0)$ the set of states for which an admissible solution to FHOCP IV exists, the following result guarantees closed-loop stability and asymptotic zero error regulation.

Thereom 8 [18]: Under A5, and for a given $y^0(\cdot) \in \Re^m$ satisfying (25), let K be such that $A_{\bar{y}^0} + B_{\bar{y}^0}K$ is asymptotically stable. Then if $X_0^{IV}(N_c, N_p, y^0)$ is nonempty, the closed-loop system (41) admits the origin as an asymptotically stable equilibrium point with output admissible set $X_0^{IV}(N_c, N_p, y^0)$.

3.6 Conclusions

In this chapter, three output feedback MPC algorithms for nonlinear systems that ensure closed-loop stability and zero-error regulation in the face of different classes of disturbance and reference signals have been presented. The first method is more general but computationally more demanding. The other two schemes guarantee zero-error regulation only for asymptotically constant references but are easily implemented in practice. The discrete-time 'separation principle' (Theorems 3 and 4) presented in Section 3.4 is remarkably general in that it guarantees closed-loop stability of the origin irrespective of the particular choice of the controller and observer (with a finite domain of attraction that is a function of those of the state feedback and the observer).

3.7 Acknowledgment

The authors acknowledge the partial financial support by MURST Project 'New techniques for the identification and adaptive control of industrial systems' (http://conpro.unipv.it/murst).

3.8 References

1 CLARKE, D.W.: 'Advances in model-based predictive control' (Oxford University Press, Oxford, UK 1994)
2 CAMACHO, E., and BORDONS, C.: 'Model predictive control in the process industry' (Springer, London, UK 1995)
3 RICHALET, J., RAULT, A., TESTUD, J.L., and PAPON, J.: 'Model predictive heuristic control: applications to industrial processes', *Automatica*, 1978, **14**, pp. 413–28
4 RICHALET, J.: 'Industrial applications of model based predictive control', *Automatica*, 1993, **29**, pp. 1251–74
5 CUTLER, C.R., and RAMAKER, B.C.: 'Dynamic matrix control – a computer control algorithm'. Automatic control conference, San Francisco, CA, USA, 1980
6 GARCIA, C.E., PRETT, D.M., and MORARI, M.: 'Model predictive control: theory and practice – a survey', *Automatica*, 1989, **25**, pp. 335–48
7 CHEN, C.C., and SHAW, L.: 'On receding horizon feedback control', *Automatica*, 1982, **18** (3), pp. 349–52
8 KEERTHI, S.S., and GILBERT, E.G.: 'Optimal, infinite-horizon feedback laws for a general class of constrained discrete-time systems', *J. Optim. Theory Appl.*, 1988, **57**, pp. 265–93
9 YANG, T.H., and POLAK, E.: 'Moving horizon control of nonlinear systems with input saturation, disturbances and plant uncertainty', *Int. J. Control*, 1993, **58**, pp. 875–903
10 DE OLIVEIRA, S.L.: 'Model predictive control (MPC) for constrained nonlinear systems'. PhD thesis, Institute of Technology, Pasadena, CA, 1996
11 MICHALSKA, H., and MAYNE, D.Q.: 'Robust receding horizon control of constrained nonlinear systems', *IEEE Trans. Automatic Control*, 1993, **38**, pp. 1512–6
12 PARISINI, T., and ZOPPOLI, R.: 'A receding-horizon regulator for nonlinear systems and a neural approximation', *Automatica*, 1995, **31**, pp. 1443–51
13 CHEN, H., and ALLGÖWER, F.: 'A quasi-infinite horizon nonlinear model predictive control scheme with guaranteed stability', *Automatica*, 1998, **34**, pp. 1205–17
14 DE NICOLAO, G., MAGNI, L., and SCATTOLINI, R.: 'Stabilising receding-horizon control of nonlinear time-varying systems', *IEEE Trans. on Automatic Control*, 1998, **43**, pp. 1030–6
15 MICHALSKA, H., and MAYNE, D.Q.: 'Moving horizon observers and observer-based control', *IEEE Trans. Automatic Control*, 1995, **40**, pp. 995–1006
16 MAGNI, L., DE NICOLAO, G., and SCATTOLINI, R.: 'Output feedback and tracking of nonlinear systems with model predictive control', *Automatica*, 2001
17 MAGNI, L., DE NICOLAO, G., and SCATTOLINI, R.: 'Output feedback receding horizon control of discrete-time nonlinear systems'. IFAC NOLCOS '98, Enschede, The Netherlands, 1998
18 DE NICOLAO, G., MAGNI, L., and SCATTOLINI, R.: 'Stabilizing predictive control of nonlinear ARX models', *Automatica*, 1997, **33**, pp. 1691–7

19 DE NICOLAO, G., MAGNI, L., and SCATTOLINI, R.: 'Tracking of nonlinear systems via model based predictive control' ADCHEM 2000, International Symposium on *Advanced control of chemical processes*, Pisa, Italy, 14–16 June 2000
20 ISIDORI, A.: 'Nonlinear control systems' (Springer-Verlag London Limited, 3rd edn, 1995)
21 KWON, W.H., and PEARSON, A.E.: 'On feedback stabilization of time-varying discrete-linear systems', *IEEE Trans. Automatic Control*, 1978, **23**, pp. 479–81
22 MAYNE, D.Q., and MICHALSKA, H.: 'Receding horizon control of nonlinear systems', *IEEE Trans. Automatic Control*, 1990, **35**, pp. 814–24
23 MEADOWS, E.S., HENSON, M.A., EATON, J.W., and RAWLINGS, J.B.: 'Receding horizon control and discontinuous state feedback stabilization', *Int. J. Control*, 1995, **62**, pp. 1217–29
24 SCOKAERT, P.O.M., RAWLINGS, J.B., and MEADOWS, E.S.: 'Discrete-time stability with perturbations: application to model predictive control', *Automatica*, 1997, **33**, pp. 463–70
25 GILBERT, E.G., and TAN, K.T.: 'Linear systems with state and control constraints: the theory and application of maximal output admissible sets', *IEEE Trans. Automatic Control*, 1991, **36**, pp. 1008–20
26 SCOKAERT, P.O.M., MAYNE, D.Q., and RAWLINGS, J.B.: 'Suboptimal model predictive control (feasibility implies stability)', *IEEE Trans. Automatic Control*, 1999, **44**, pp. 648–54
27 MAGNI, L.: 'Nonlinear receding horizon control: theory and application'. PhD thesis, Dipartimento di Informatica e Sistemistica, Università degli Studi di Pavia, 1998
28 ZHENG, A.: 'Model predictive control: Is QP necessary?'. AIChE Annual Meeting, Chicago, IL, USA, 1996
29 SZNAIER, M., and DAMBORG, M.: 'Suboptimal control of linear systems with state and control inequality constraints'. Proceedings of 26th Conference on Decision and Control, 1987, pp. 761–2
30 SZNAIER, M., and DAMBORG, M.J.: 'Control of constrained discrete time linear systems using quantized controls', *Automatica*, 1989, **25**, pp. 623–8
31 RAWLINGS, J.B., and MUSKE, K.R.: 'The stability of constrained receding horizon control', *IEEE Trans. Automatic Control*, 1993, **38**, pp. 1512–6
32 SCOKAERT, P.O.M., and RAWLINGS, J.B.: 'Infinite horizon linear quadratic control with constraints'. IFAC 13th Triennal World Congress, 1996, pp. M 109–14
33 MAGNI, L., DE NICOLAO, G., MAGNANI, L., and SCATTOLINI, R.: 'A stabilizing model-based predictive control for nonlinear systems', *Automatica*, 2001
34 DE NICOLAO, G., MAGNI, L., and SCATTOLINI, R.: 'Stability and robustness of nonlinear receding-horizon control', in ALLGÖWER, F., and ZHENG, A. (Eds): 'Nonlinear model predictive control', *Progress in Systems and Control Theory* (Birkhauser Verlag, 2000), pp. 3–22
35 MAYNE, D.Q.: 'Nonlinear model predictive control: challenges and opportunities', in 'NMPC – assessment and future directions' (Birkhauser Verlag, 1999)

36 ALLGÖWER, F., BADGWELL, T.A., QIN, J.S., RAWLINGS, J.B., and WRIGHT, S.: 'Nonlinear predictive control and moving horizon estimation – an introductory overview', in FRANK, P.M., (Ed.): 'Advances in control', (Springer, 1999), pp. 391–449
37 VIDYASAGAR, M.: 'Nonlinear system theory' (Prentice-Hall, Englewood Cliffs, NJ, 2nd edn, 1993)
38 VIDYASAGAR, M.: 'On the stabilization of nonlinear systems using state detection', *IEEE Trans. Automatic Control*, 1980, **25**, pp. 504–9
39 ROBERTSON, D.G., LEE, J.H., and RAWLINGS, J.B.: 'A moving horizon-based approach to least squares estimation', *AIChE Journal*, 1996, **42**, pp. 2209–24
40 ALESSANDRI, A., PARISINI, T., and ZOPPOLI, R.: 'Neural approximations for nonlinear finite-memory state estimation', *International Journal of Control*, 1997, **67** (2), pp. 275–302
41 ALESSANDRI, A., BAGLIETTO, M., PARISINI, T., and ZOPPOLI, R.: 'A neural state estimator with bounded errors for nonlinear systems', *IEEE Trans. Automatic Control*, 1999, **44** (11), pp. 2028–42
42 SONG, Y., and GRIZZLE, J.W.: 'The extended Kalman filter as a local asymptotic observer for nonlinear discrete-time systems', *J. Math. Syst. Estim. Control*, 1995, **5**, pp. 59–78
43 ISIDORI, A., and BYRNES, C.I.: 'Output regulation of nonlinear systems', *IEEE Trans. Automatic Control*, 1990, **35**, pp. 131–40
44 LEVIN, A.U., and NARENDRA, K.S.: 'Control of nonlinear dynamical systems using neural networks – Part II: Observability, identification, and control', *IEEE Trans. Neural Netw.*, 1996, **7**, pp. 30–42
45 CHEN, S., BILLINGS, S.A., and LUO, W.: 'Orthogonal least squares methods and their application to nonlinear system identification', *Int. Journal of Control*, 1989, **50**, pp. 1873–96
46 HABER, R., and UNBEHAUEN, H.: 'Structure identification of nonlinear dynamic systems – a survey on input/output approaches', *Automatica*, 1990, **26**, pp. 651–77
47 LEONTARITIS, I.J., and BILLINGS, S.A.: 'Input-output parametric models for nonlinear systems. Part I: Deterministic non-linear systems', *Int. Journal Control*, 1985, **41**, pp. 303–28
48 LEVIN, A.U., and NARENDRA, K.S.: 'Control of nonlinear dynamical systems using neural networks: controllability and stabilization', *IEEE Trans. Neural Netw.*, 1993, **4**, pp. 192–206

Chapter 4

Model predictive control of nonlinear parameter varying systems via receding horizon control Lyapunov functions

Mario Sznaier and James Cloutier

Abstract

The problem of rendering the origin an asymptotically stable equilibrium point of a nonlinear system while, at the same time, optimising some measure of performance has been the object of much attention in the past few years. In contrast to the case of linear systems where several optimal synthesis techniques (such as H_∞, H_2 *and* ℓ^1*) are well established, their nonlinear counterparts are just starting to emerge. Moreover, in most cases these tools lead to partial differential equations that are difficult to solve. In this chapter we propose a suboptimal regulator for nonlinear parameter varying, control affine systems based upon the combination of model predictive and control Lyapunov function techniques. The main result of the chapter shows that this controller is nearly optimal provided that a certain finite horizon problem can be solved on-line. Additional results include: (a) sufficient conditions guaranteeing closed-loop stability even in cases where there is not enough computational power available to solve this optimisation on-line; and (b) an analysis of the suboptimality level of the proposed method.*

4.1 Introduction

A large number of control problems involve designing a controller capable of rendering some point into an asymptotically stable equilibrium point of a given time invariant system while, simultaneously, optimising some performance index. In the case of linear dynamics this problem has been thoroughly explored during the past decade, leading to powerful formalisms such as μ-synthesis and ℓ^1optimal control theory that have been successfully employed to solve some hard practical problems.

In the case of nonlinear dynamics, popular design techniques include Jacobian linearisation (JL) [1], feedback linearisation (FL) [1], the use of control Lyapunov functions (CLF) [2, 3] and recursive backstepping [1]. While these methods provide powerful tools for designing stabilising controllers, performance of the resulting closed-loop systems can vary widely, as we illustrate in the sequel with the problem of controlling a thrust vectored aircraft. A simplified planar model of the system is shown in Figure 4.1, with the corresponding dynamics given by (see References 4 and 5 for details):

$$\begin{bmatrix} \ddot{x} \\ \ddot{y} \\ \ddot{\theta} \end{bmatrix} = \begin{bmatrix} -g\sin\theta \\ g(\cos\theta - 1) \\ 0 \end{bmatrix} + \begin{bmatrix} \frac{\cos\theta}{m} & -\frac{\sin\theta}{m} \\ \frac{\sin\theta}{m} & \frac{\cos\theta}{m} \\ \frac{r}{J} & 0 \end{bmatrix} \begin{bmatrix} u_1 \\ u_2 \end{bmatrix} \tag{1}$$

where x, y and θ denote horizontal, vertical and angular position, respectively, and where u_1 and $u_2 = \tilde{u}_2 + mg$ are the control inputs. (Following Reference 4 the control u_2 has been shifted to compensate for gravity.)

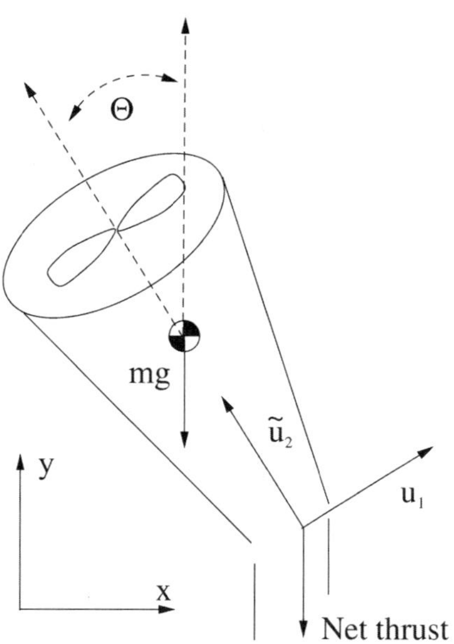

Figure 4.1 Simplified model of a thrust vectored aircraft

Assume that the goal is to drive the system to the origin, while minimising a performance index of the form

$$J(x_0, u) = \int_0^{\infty} [\xi^T Q \xi + u^T R u] dt \tag{2}$$

$$\xi(0) = [0 \quad 0 \quad 0 \quad 12.5 \quad 0 \quad 0]^T \tag{3}$$

$$Q = \text{diag}\{[5 \quad 5 \quad 1 \quad 1 \quad 1 \quad 5]\}, \quad R = I_{2\times 2} \tag{4}$$

corresponding to the following choice of state variables: $\xi = [x \quad y \quad \theta \quad \dot{x} \quad \dot{y} \quad \dot{\theta}]^T$.

Table 4.1 compares the performance achieved by several commonly used nonlinear control design methods. As shown there, in this case the performance of the CLF and JL controllers is worse, by an order of magnitude, than the optimal cost (obtained by offline optimisation using a conjugate gradient algorithm). Indeed, a recent workshop on nonlinear control [5] has shown that, while the methods mentioned above can recover the optimal control under certain conditions, in general there are no guarantees on the performance of the resulting system.

As an alternative, nonlinear counterparts of H_∞ [6, 7] and ℓ^1 [8] have recently started to emerge. While these theories can guarantee optimality (at least in a certain sense), from a practical standpoint they suffer from the fact that they lead to Hamilton-Jacobi-Isaacs type partial-differential equations that are hard to solve, except in some restrictive, low-dimensional cases.

Given these practical difficulties, during the past few years there has been an increased interest in extending receding horizon (RH) techniques to nonlinear plants. These techniques are appealing since they allow for explicitly handling constraints and guarantee optimality in some sense. Moreover, since the optimisation is carried only along the present trajectory of the system (ie. 'locally') the resulting computational complexity is far less than that associated with finding the true global optimal control (a task that entails solving a Hamilton-Jacobi type equation). However, in contrast with the linear case (where global stability has been established [9–11]), for nonlinear plants only local stability results are available [12]. Several modified nonlinear RH formulations addressing this problem have been proposed, mostly based on the use of

Table 4.1 Comparison of different methods for the thrust vectored aircraft example

Method	*Cost*
Exact	1115

additional constraints or a terminal penalty. For instance, Reference 13 uses dual mode control, where an explicit optimisation is used to drive the system to a neighbourhood of the origin, where a locally stabilising linear control law is used. Magni [14] uses a terminal penalty obtained assuming that a linear control law will be used after the optimisation horizon T. Finally, Reference 15 achieves stability by enforcing an additional state constraint. However, while these approaches guarantee closed-loop stability, they may do so at the expense of performance.

In this chapter we propose an alternative controller for suboptimal regulation of nonlinear, parameter varying, control affine systems, based upon the combination of receding horizon and control Lyapunov functions ideas. This approach follows in the spirit of a similar controller successfully used in the case of constrained linear systems [9, 16, 17], in which receding horizon control steers the system to an invariant neighbourhood of the origin where a stabilising controller is available. In the first part of the chapter we show that combining these ideas with a suitable finite-horizon approximation of the performance index leads to globally stabilising, nearly optimal controllers, provided that enough computational power is available to solve on-line an optimisation problem. In the second part of the chapter we show how to modify these controllers to guarantee global stability in the face of computational time constraints. Additional results include an analysis of the suboptimality of the proposed method and show that, if an approximate solution to the problem is known in a set containing the origin, then our controller yields an extension of this solution with the same suboptimality level. Finally, we show that in the limit as the optimisation horizon $T \rightarrow 0$, the method reduces to the well known inverse optimal controller of Freeman and Kokotovic [18].

4.2 Preliminaries

4.2.1 Notation and definitions

In the sequel we consider the following class of control-affine nonlinear parameter varying (NLPV) systems:

$$\dot{x} = f[x, \rho(t)] + g[x, \rho(t)]u \tag{5}$$

where $x \in \mathbf{R}^n$ and $u \in \mathbf{R}^m$ represent the state and control variables, the vector fields $f(\cdot,\cdot)$ and $g(\cdot,\cdot)$ are known C^1 functions, and where $\rho \in \mathbf{R}^{n_\rho}$ denotes a vector of time-varying parameters, unknown *a priori*, but available to the controller in real time. Further, we will assume that at all times $\rho \in \mathbf{P} \subset \mathbf{R}^{n_\rho}$, where $\mathbf{P}$

is a given compact set, and that the set of admissible parameter trajectories is given by

$$F_\nu := \{\rho \in C^1(\mathbf{R}, \mathbf{R}^{n_\rho}) : \rho(t) \in \mathbf{P}, \quad \underline{\nu}_i \leq \dot{\rho}_i(t) \leq \bar{\nu}_i, \quad i = 1, 2, \ldots n_\rho, \quad \forall t \in \mathbf{R}_+\} \quad (6)$$

where $\underline{\nu}_I$ and $\bar{\nu}_I$ are given numbers.

Definition 1: *A positive definite C^1 function $V : \mathbf{R}^n \times \mathbf{R} \rightarrow \mathbf{R}$ is a control Lyapunov function (CLF) for the system (5) if it is radially unbounded in x and*

$$\inf_u \left[L_f V(x, \rho) + L_g V(x, \rho) + \frac{\partial V}{\partial \rho} \dot{\rho} \right] \leq -\sigma(x) > 0, \quad \forall x \neq 0 \text{ and all } \rho \in F_\nu \quad (7)$$

where $\sigma(\cdot)$ is a positive definite function, and where $L_h V(x, \rho) = \frac{\partial V}{\partial x} h(x, \rho)$ denotes the Lie derivative of V along h.

4.2.2 Quadratic regulator problem for NLPV systems

Consider the NLPV system (5). In the sequel we consider the following problem. Given an initial condition x_0 and an initial value of the parameter ρ_o, find a parameter dependent state-feedback control law $u[x(t), \rho(t)]$ that minimises the following performance index:

$$J(x_0, \rho_0, u) = \sup_{\rho \in \mathbf{F}_\nu, \rho(0) = \rho_0} \int_0^\infty [x^T Q(x, \rho) x + u^T R(x, \rho) u] \, dt, \quad x(0) = x_0 \quad (8)$$

where $Q(\cdot)$ and $R(\cdot)$ are C^1, positive definite matrices (this condition can be relaxed to $Q(x, \rho) \geq 0$). In the sequel, for simplicity, the explicit dependence of matrices on x and ρ will be omitted, when it is clear from the context.

By using Pontryagin's principle it can be shown that solving this problem is equivalent to solving the following Hamilton-Jacobi-Bellman type partial differential equation:

$$0 = \frac{\partial V}{\partial x} f - \frac{1}{4} \frac{\partial V}{\partial x} g R^{-1} g^T \frac{\partial V^T}{\partial x} + \frac{\partial V}{\partial t} + x^T Q x + \max_{\underline{\nu}_i \leq \nu \leq \bar{\nu}_i} \sum_{i=1}^{n_\rho} \frac{\partial V}{\partial \rho_i} \nu_i \quad (9)$$

subject to $V(0, \rho) = 0$

If this equation admits a C^1 nonnegative solution V, then the optimal control is given by $u(x, \rho) = -\frac{1}{2} R^{-1} g^T \frac{\partial V^T}{\partial x}$ and $V(x, \rho)$ is the corresponding optimal cost (or storage function), i.e.

$$V(x, \rho) = \min_u \sup_{\rho \in \mathbf{F}_\nu} \int_0^\infty [x^T Q x + u^T R u] \, dt$$

4.3 Equivalent finite horizon regulation problem

Unfortunately, the complexity of (9) prevents its solution, except in some very simple, low dimensional cases. To solve this difficulty, motivated by the work in References 16 and 19, in this section we introduce a finite horizon approximation to the nonlinear regulation problem.

Lemma 1: Consider a compact set S containing the origin in its interior and assume that the optimal storage function $V(x, \rho)$ is known for all $x \in S$, $\rho \in \mathbf{P}$. Let $v = \min_{x \in \partial S} \min_{\rho \in P} V(x, \rho)$ where ∂S denotes the boundary of S. Finally, define the set $S_v = \{x : \sup_{\rho \in P} V(x, \rho) \leq c\}$. Consider the following two optimisation problems:

$$\min_{u} \sup_{\rho \in \mathbf{F}_\nu, \rho(t) = \rho_0} \left\{ J(x_0, u, \rho_0) = \int_t^\infty [x^T Q x + u^T R u] d\tau \right\} \tag{10}$$

$$\min_{u} \sup_{\rho \in \mathbf{F}_\nu, \rho(t) = \rho_0} \left\{ J_T(x_0, u, \rho_0) = \int_t^T [x^T Q x + u^T R u] d\tau + V[x(T), \rho(T)] \right\} \tag{11}$$

subject to (5) with $x(t) = x_0$. Then an optimal solution of problem (11) is also optimal for (10) in the interval $[t, T]$ provided that $x(T) \in S_v$.

Proof: If $x_0 \in S_v$ the proof follows immediately from the facts that, for all admissible parameter trajectories, S_v is positively invariant and that $V(x, \rho)$ is the optimal return function there. If $x \notin S_v$, consider the following free terminal time problem:

$$J^0(x, \rho, t) = \min_{u} \sup_{\rho \in \mathbf{F}_\nu} \left\{ V[x(t_f), \rho(t_f)] + \int_t^{t_f} [x^T Q x + u^T R u] d\tau \right\}$$
$$\text{subject to } x(t_f) \in S_v \tag{12}$$

Let x^0, u^0, ρ^0 denote the optimal trajectory. It can be easily seen that the optimal return function satisfies

$$0 = \frac{\partial J}{\partial t} + \frac{\partial J}{\partial x} f(x, \rho) - \frac{1}{4} \frac{\partial J}{\partial x} g R^{-1} g^T \frac{\partial V^T}{\partial x} + x^T Q x + \max_{\underline{\nu}_i \leq \nu \leq \bar{\nu}_i} \sum_{i=1}^{n_\rho} \frac{\partial J}{\partial \rho_i} \nu_i \tag{13}$$

with boundary condition $J(x, \rho, t) = V(x, \rho)$ for $x \in S_v$. Clearly this equation admits as solution $J(x, \rho, t) = V(x, \rho)$. Thus problems (10) and (12) are equivalent. To establish the claim we will show that an optimal solution u^0 of (11) is also optimal for (12) (and thus (10)), provided that $x^0(T) \in S_v$. To this effect note that the Euler-Lagrange optimality conditions for problems (11) and (12) are identical, except for the additional transversality condition $H[u^0, x^0(t_f), \lambda^0(t_f), \mu^0(t_f)] = 0$ that appears in the latter, where $\lambda(t)$ and $\mu(t)$ denote the co-states associated with the states and parameters, respectively. The boundary conditions for these co-states in problem (11) are given by

$$\lambda^0(T) = \left.\frac{\partial V}{\partial x}\right|_{x(T),\rho(T)}, \quad \mu^0(T) = \left.\frac{\partial V}{\partial \rho}\right|_{x(T),\rho(T)} \tag{14}$$

Since $x(T) \in S_v$ it follows that $x^0(T), u^0(T), \lambda^0(T), \mu^0(T)$ satisfy the HJB eqn (9), or equivalently $H[u^0, x^0(T), \lambda^0(T), \mu^0(T)] = 0$. Thus, an optimal solution of (11) is also optimal for (12).

This lemma shows that if a solution to the HJB equation (9) is known in a neighbourhood of the origin, then it can be extended via an explicit finite horizon optimisation, well suited for an on-line implementation. This suggests the following RH type control law.

Algorithm 1:

0. *Data: The region S_v, the function $V(x,\rho)$ for all $x \in S_v$ a sampling interval δT.*
1. *If $x(t) \in S_v, u = -\frac{1}{2}R^{-1}g^T(x,\rho)\frac{\partial V^T}{\partial x}(x,\rho)$.*
2. *If $x(t) \notin S_{v_y}$ then solve a sequence of optimisation problems of the form (11) with increasing values of T until a solution such that $x(T) \in S_v$ is found. Use the corresponding control law $u(t)$ in the interval $[t, t+\delta T]$.*

From the results above it is clear that the resulting control law is globally optimal and thus globally stabilising. However, the computational complexity associated with finding $V(x,\rho)$ (even only in the region S_v) may preclude the use of this control law in many practical cases. Thus, it is of interest to consider a control law where an approximation $\Psi(x,\rho)$ (rather than $V(x,\rho)$) is used. To this effect consider a compact set S containing the origin in its interior and let $\Psi : S \times \mathbf{P} \to \mathbf{R}_+, \Psi \in C^1(\mathbf{R}^{n_x} \times \mathbf{R}^{n_\rho}, \mathbf{R})$ be a control Lyapunov function for system (5). Finally, let $c = \min_{x \in \partial S} \min_{\rho \in P} \Psi(x,\rho)$ and define the set $S_\Psi \subseteq S$ by $S_\Psi = \{x : \max_{\rho \in P} \Psi(x,\rho) \le c\}$. Consider the modified control law as follows.

Algorithm 2:

0. *Data: a CLF $\Psi(x,\rho)$ the region S_Ψ, a sampling interval δT, a positive definite function $\sigma(\cdot)$.*
1. *If $x(t) \in S_\Psi$ then*

$$u_\Psi(x,\rho) = \arg\min_u \times \left\{ \|u\| : \sup_{\underline{\nu}_i \le \nu \le \bar{\nu}_i} \frac{\partial \Psi}{\partial x}[f(x,\rho) + g(x,\rho)u] + \sum_i \frac{\partial \Psi}{\partial \rho_i}\nu_i \le -\sigma(x) < 0 \right\}$$

2. If $x(t) \notin S_\Psi$ then consider an increasing sequence T_i. Let

$$u^*_{T_i} = \arg\min_u \sup_{\rho \in \mathbf{F}_\nu} \left\{ \int_t^{T_i} [x^T Qx + u^T Ru] d\tau + \Psi[x(T_i), \rho(T_i)] \right\}$$

Denote by $x^*(\cdot)$ the corresponding optimal trajectory and define:

$$J(x_0, \rho_0, T_i) = \sup_{\rho \in \mathbf{F}_\nu} \left\{ \int_t^{T_i} [x^{*T} Qx^* + u^{*T} Ru^*] d\tau + \Psi[x^*(T_i), \rho(T_i)] \right\}$$

$$T(x, \rho_0) = \arg\min_T \{ J(x, \rho_0, T_i) : x^*(T_i) \in S_\Psi \text{ for all } \rho(\cdot) \in \mathbf{F}_\nu, \rho(t) = \rho_0 \} \quad (15)$$

Then $u_\Psi(x, \rho) := u^*_{T(x)}(\tau), \tau \in [t, t + \delta T]$.

(Note that from LaSalle-Yoshizawa's Theorem [1] we have that $\lim_{T \to \infty} x^T Qx = 0$. Hence $T(x_0, \rho_0)$ is finite.)

Thereom 1: Assume that $Q(x, \rho) \geq \sigma_m I > 0$. Then the control law u_Ψ generated by Algorithm 2 renders the origin a globally asymptotically stable equilibrium point of (5).

Proof: Consider first an initial condition $x_0 \notin S_\Psi$, an initial value of the parameter ρ_0, and the corresponding optimal control law $u^*(\cdot)$ and trajectory $x^*(\cdot)$. Let

$$J(x_0, \rho_0) = \sup_{\rho \in \mathbf{F}_\nu, \rho(t) = \rho_0} \times \left\{ \int_t^{T(x_0, \rho_0)} [x^{*T} Qx^* + u^{*T} Ru^*] d\tau + \Psi\{x^*[T(x_0, \rho_0)], \rho[T(x_0, \rho_0)]\} \right\}$$

and define $x_1 := x^*(t + dt)$ (with $dt > 0 >$ small enough so that $x^*(t + dt) \notin S_\Psi$) and $\rho_1 := \rho(t + dt)$, where $\rho(\cdot)$ denotes any admissible parameter trajectory starting at ρ_0. Since $x^*(\tau), \tau \in [t + dt, T]$ is also a feasible trajectory starting from x_1 we have that

$$\begin{aligned} J(x_1, \rho_1) &= \inf_u \sup_{\rho \in \mathbf{F}_\nu, \rho(t+dt) = \rho_1} \left\{ \int_{t+dt}^{T(x_1, \rho_1)} [x^{*T} Qx^* + u^{*T} Ru^*] d\tau + \Psi\{x^*[T(x_1, \rho_1)], \rho[T(x_1, \rho_1)]\} \right\} \\ &\leq \sup_{\rho \in \mathbf{F}_\nu, \rho(t+dt) = \rho_1} \left\{ \int_{t+dt}^{T(x_0, \rho_0)} [x^{*T} Qx^* + u^{*T} Ru^*] d\tau + \Psi\{x^*[T(x_0, \rho_0)], \rho[T(x_0, \rho_0)]\} \right\} \\ &\leq \sup_{\rho \in \mathbf{F}_\nu, \rho(t) = \rho_0} \left\{ \int_t^{T(x_0, \rho_0)} [x^{*T} Qx^* + u^{*T} Ru^*] d\tau + \Psi\{x^*[T(x_0, \rho_0)], \rho[T(x_0, \rho_0)]\} \right\} \\ &\quad - \int_t^{t+dt} [x^{*T} Qx^* + u^{*T} Ru^*] d\tau = J(x_0, \rho_0) - \int_t^{t+dt} [x^{*T} Qx^* + u^{*T} Ru^*] d\tau. \end{aligned} \quad (16)$$

Thus, for any admissible parameter trajectory,

$$\dot{J} = \lim_{dt\to 0} \frac{J[x(t+dt), \rho(t+dt)] - J[x(t), \rho(t)]}{dt} \leq -x^T(t)Qx(t) \leq -\sigma_m \cdot \|x\|^2 < 0$$

Since $J(x, \rho) > 0$ and $\dot{J}(x, \rho) < 0$ for all $x \notin S_\Psi$, it follows that all trajectories reach the set S_Ψ in finite time. Asymptotic stability now follows from the facts that S_Ψ is invariant with respect to u_Ψ (i.e. trajectories starting in the set never leave it) and that $\Psi(x, \rho)$ is a CLF there.

4.4 Modified receding horizon controller

In the last section we outlined a receding horizon type law, that under certain conditions globally stabilises system (5). While most of these conditions are rather mild (essentially equivalent to the existence of a CLF), the requirement that T should be large enough so that $x(T) \in S_\Psi$ could pose a problem, especially in cases where the system has fast dynamics. Thus, it is of interest to consider the following modified control law where both an approximation $\Psi(x, \rho)$ (rather than $V(x, \rho)$) and a fixed horizon T are used.

Algorithm 3:

0. *Data: a CLF $\Psi(x)$, an invariant region S_Ψ such that $0 \in int(S_\Psi)$, a horizon T.*
1. *If $x(t) \in S_\Psi$ then*

$$u_\Psi(x, \rho) = \arg\min_u$$
$$\times \left\{ \|u\| : \sup_{\underline{\nu}_i \leq \nu \leq \bar{\nu}_i} \frac{\partial \Psi}{\partial x}[f(x, \rho) + g(x, \rho)u] + \sum_i \frac{\partial \Psi}{\partial \rho_i}\nu_i \leq -\sigma(x) < 0 \right\}$$

2. *If $x(t) \notin S_\Psi$ then $u_\Psi(x, \rho) = u(t)$ where $u(\lambda), \lambda \in [t, t+T]$ is given by:*

$$u = \arg\min_u \sup_{\rho \in \mathbf{F}_\nu} \left\{ \int_t^{T+t} [x^T Qx + u^T Ru] d\tau + \Psi[x(T+t), \rho(T+t)] \right\} \tag{17}$$

subject to:

$$0 \geq x^T(t+T)Qx(t+T) + \min_w \sup_{\rho \in \mathbf{F}_\nu} \left\{ w^T Rw + \left.\frac{d\Psi}{dt}\right|_{x(t+T)} \right\} - u^T(t)Ru(t) \tag{18}$$

Note that, inside the region S_Ψ, the control action is a generalisation to the NLPV case of the pointwise-minimum norm controller proposed in Reference 18.

Proceeding as in Reference 20 it can be shown that if the CLF Ψ has the same level sets as the optimal value function V inside S_Ψ, then u_Ψ is indeed the optimal solution for the original problem. Thus, the algorithm allows for 'switching-off' the on-line optimisation when inside a region where the optimal storage function (or a good approximation) is known.

Thereom 2: The control law u_Ψ generated by Algorithm 3 has the following properties:

1. It renders the origin a globally asymptotically stable equilibrium point of (5).
2. It coincides with the globally optimal control law when $\Psi(x,\rho) = V(x,\rho)$, where $V(x,\rho)$ denotes the optimal storage function obtained by solving (9).

Proof: To prove stability, proceeding as in Theorem 1, consider first an initial condition $x \notin S_\Psi$. Denote by u^*, x^* the optimal control and associated trajectory, respectively. Then

$$\begin{aligned} J[x(t+dt),\rho(t+dt)] &= \min_u \sup_\rho \left\{ \int_{t+dt}^{T+t+dt} [x^T Qx + u^T Ru]d\tau + \Psi[x(T+t+dt),\rho(T+t+dt)] \right\} \\ &\le \sup_\rho \int_{t+dt}^{T+t} [x^{*T} Qx^* + u^{*T} Ru^*]d\tau + \Psi[x^*(T+t+dt),\rho(T+t+dt)] \\ &\quad + \min_w \sup_\rho \{x^{*T}(T+t)Qx^*(T+t) + w^T Rw + \dot{\Psi}[x^*(T+t),\rho]\}dt \\ &\le J[x(t),\rho(t)] - [x^{*T}(t)Qx^*(t) + u^{*T}(t)Ru^*(t)]dt \\ &\quad + \min_w \sup_\rho \{x^{*T}(T+t)Qx^*(T+t) + w^T Rw + \dot{\Psi}[x^*(T+t),\rho]\}dt. \end{aligned} \tag{19}$$

Therefore, if (18) holds then we have that, for all admissible parameter trajectories,

$$\dot{J} = \lim_{dt\to 0} \frac{J[x(t+dt),\rho(t+dt)] - J[x(t),\rho(t)]}{dt} \le -x^T(t)Qx(t) \le -\sigma_m \cdot \|x\|^2 < 0 \tag{20}$$

where σ_m denotes the minimum singular value of Q. Hence the trajectories starting outside S_Ψ reach this set in a finite time. As in the proof of Theorem 1, once there, asymptotic stability is guaranteed by the parameter dependent control Lyapunov function $\Psi(x,\rho)$.

To prove item 2 note that if $\Psi(x,\rho) = V(x,\rho)$ then from the Hamilton Jacobi equation (9) we have that

$$x^T(t+T)Qx(t+T) + \min_w \sup_{\rho\in \mathbf{F}_\nu} \left\{ w^T Rw + \dot{\Psi}\big|_{(t+T)} \right\} = 0 \tag{21}$$

Thus the constraint (18) is redundant and the proof follows immediately from Lemma 1.

Finally, before closing this section we consider a modified control law that takes into account the sample and hold nature of receding horizon implementations.

Algorithm 4:

0. *Data: a CLF* $\Psi(x)$ *an invariant region* S_Ψ *such that* $0 \in \text{int}(S_\Psi)$ *, a horizon* T*, a sampling interval* δT.
1. *If* $x(t) \in S_\Psi$,

$$u_\Psi(x, \rho) = \arg\min_u$$

$$\times \left\{ \|u\| : \sup_{\underline{\nu}_i \le \nu \le \bar{\nu}_i} \frac{\partial \Psi}{\partial x}[f(x,\rho) + g(x,\rho)u] + \sum\nolimits_i \frac{\partial \Psi}{\partial \rho_i}\nu_i \le -\sigma(x) < 0 \right\}$$

2. *If* $x(t) \notin S_\Psi$ *then* $u_\Psi(x,\rho) = u^*(\tau)$ *where* $u^*(\cdot)$ *is given by*

$$u^* = \arg\min_u \sup_{\rho \in \mathbf{F}_\nu} \left\{ \int_t^{T+t} [x^T Q x + u^T R u] d\tau + \Psi[x(T+t), \rho(T+t)] \right\} \tag{22}$$

subject to:

$$\begin{aligned} 0 \ge\ & x^T(t+T+\tau)Qx(t+T+\tau) \\ & + \min_u \sup_{\rho \in \mathbf{F}_\nu} \left\{ u^T(t+T+\tau)Ru(t+T+\tau) + \dot{\Psi}\big|_{(t+T+\tau)} \right\} - u^T(t)Ru(t) \end{aligned} \tag{23}$$

for all $0 \le \tau \le \delta T$

Lemma 2: The control law u^* renders the origin a globally stable equilibrium point of (5). Moreover, it coincides with the globally optimal control law when $\Psi(x,\rho) = V(x,\rho)$ and $\delta T \to 0$.

Proof: The proof, which is omitted for space reasons, follows along the lines of the proof of Theorem 2.

4.5 Selecting suitable CLFs

In principle, any of the methods available in the literature for finding CLFs such as feedback linearisation and backstepping (see, for instance, Reference 21) can be used to find the function $\Psi(\cdot)$. Alternatively, if a stabilising linear static feedback control law $u_K = Kx$ is known, then, following Reference 14 a suitable CLF is given by

$$\Psi(x,\rho) = \sup_{\rho \in \mathbf{F}_\nu} \int_t^\infty x^T(Q + K^T R K)x d\tau$$

However, as we show next, in some cases of practical interest, specific families of CLFs are readily available that reduce the degree of suboptimality incurred by the algorithm.

4.5.1 Autonomous systems

Consider the case of autonomous nonlinear systems, i.e. where $f = f(x)$ and $g = g(x)$ in (5). As we show in the sequel, in this case the suboptimality level incurred by the proposed algorithm is roughly similar to the difference between the CLF $\Psi(\cdot, \cdot)$ and the actual value function $V(\cdot, \cdot)$.

Thereom 3: Let $\Psi : \mathbf{R}^n \to \mathbf{R}_+$ be a positive definite, radially unbounded function and consider the following optimisation problem:

$$J_\Psi(x,t) = \min_u \int_t^T [x^T Q x + u^T R u] d\tau + \Psi[x(T)] \tag{24}$$

subject to (5). Then

$$J_\Psi(x,t) - V[x(t)] = \Psi[x(T)] - V[x(t)] + O\left[\left|\left| \frac{\partial e}{\partial x}\Big|_{x(T)} \right|\right|^2 \right] + O(dt^2) \tag{25}$$

where $e(x,t) := J_\Psi(x,t) - V(x)$ denotes the approximation error.

Proof: By considering the Hamilton Jacobi equations for J_Ψ and V it can be easily shown that $e(t,x)$ satisfies the following equation:

$$0 = \frac{\partial e}{\partial t} + \frac{\partial e}{\partial x}\left(f - \frac{1}{2} g R^{-1} g^T \frac{\partial J_\Psi}{\partial x}^T \right) + \frac{1}{4}\frac{\partial e}{\partial x} g R^{-1} g^T \frac{\partial e}{\partial x}^T \tag{26}$$

By exploiting the fact that the optimal control law for (24) is given by

$$u_\Psi = -\frac{1}{2} R^{-1} g^T \frac{\partial J_\Psi}{\partial x}^T \tag{27}$$

(26) can be rewritten as

$$\begin{aligned} 0 &= \frac{\partial e}{\partial t}^T + \frac{\partial e}{\partial x}(f - g u_\Psi) + \frac{1}{4}\frac{\partial e}{\partial x} g R^{-1} g^T \frac{\partial e}{\partial x}^T \\ &= \frac{\partial e}{\partial t} + \frac{\partial e}{\partial x}\dot{x} + \frac{1}{4}\frac{\partial e}{\partial x} g R^{-1} g^T \frac{\partial e}{\partial x}^T \\ &= \dot{e} + \frac{1}{4}\frac{\partial e}{\partial x} g R^{-1} g^T \frac{\partial e}{\partial x}^T \end{aligned} \tag{28}$$

From this last equation it follows that

$$\dot{e}(T) = -\frac{1}{4}\frac{\partial e}{\partial x} g R^{-1} g^T \frac{\partial e^T}{\partial x}\bigg|_{x(T)} \tag{29}$$

Expanding $e(t)$ in a Taylor series around $t = T$ yields

$$\begin{aligned} e(t) &= e(T) + \dot{e}(T)dt + O(dt^2) \\ &= e(T) - \frac{1}{4}\frac{\partial e}{\partial x} g R^{-1} g^T \frac{\partial e^T}{\partial x}\,|_{x(T)} dt + O(dt^2) \end{aligned} \tag{30}$$

Corollary 1: *Assume that* Ψ *is selected so that* $||\frac{\partial e[x(T),T]}{\partial x}|| \cong 0$. *Then* $J_\Psi(x,t) - V[x(t)] \cong \Psi[x(T)] - V[x(T)]$ *(to the first order in dt) along the trajectories of the system.*

The result above formalises the intuitively appealing fact that in order to improve performance, the CLF $\Psi(x)$ should be selected 'close' to V. Next we briefly discuss an approach to generate such functions. This approach is motivated by the empirically observed success of the State Dependent Riccati Equation (SDRE) method, briefly covered in the Appendix.

From Lemma 4 in the Appendix, it follows that $\Psi(x) = x^T P(x)x$, where $P(x)$ denotes the solution to the SDRE, is a CLF in a neighbourhood of the origin. Moreover, since the control law u_{SDRE} is locally stabilising, it can be easily shown that there exists T_0 (possibly depending on the initial condition) such for all $T > T_0$ the constraints (18) are feasible. It follows that $\Psi(x) = x^T P(x)x$ is a suitable choice for the terminal penalty. Moreover from the properties of the SDRE method (see the Appendix) it follows that with this choice, the control law satisfies all the necessary conditions for optimality as $O(||x(t+T)||^2)$. Thus, we will expect that Algorithm 3 using $\Psi(x) = x^T P(x)x$ will generate a nearly optimal control law, even when T is relatively small. In Section 4.8 we show that this is indeed the case with two examples.

4.5.2 Linear parameter varying systems

Consider now the case where the dynamics (5) are linear in the state, i.e.

$$\dot{x} = A[\rho(t)]x + B[\rho(t)]u \tag{31}$$

This is the case of LPV systems, that has been the object of much attention in the past few years [22], as a vehicle to formalise the concept of gain scheduling. In this case it can be shown that a parameter dependent state feedback controller with guaranteed performance can be synthesised by solving the following convex optimisation problem (see Reference 23 for details):

$$\min_{X(\rho)>0,Z} \quad \text{trace}(Z)$$

subject to

$$\begin{bmatrix} -\sum_{i=1}^{n_\pi} \underline{\nu}_i \frac{\partial X(\rho)}{\delta \rho_i} + A(\rho)X(\rho) + X(\rho)A^T(\rho) - B(\rho)R^{-1}B^T(\rho) & X(\rho) \\ X(\rho) & -Q \end{bmatrix} < 0$$
$$\begin{bmatrix} -\sum_{i=1}^{n_\pi} \bar{\nu}_i \frac{\partial X(\rho)}{\delta \rho_i} + A(\rho)X(\rho) + X(\rho)A^T(\rho) - B(\rho)R^{-1}B^T(\rho) & X(\rho) \\ X(\rho) & -Q \end{bmatrix} < 0 \quad (32)$$
$$\begin{bmatrix} Z & I \\ I & X(\rho) \end{bmatrix} > 0$$

for all $\rho \in \mathbf{P}$. The corresponding control action is given by

$$u = -\frac{1}{2}R^{-1}B^T(\rho)X^{-1}(\rho)x \quad (33)$$

and the corresponding cost is bounded by

$$J(x_0, \rho_0) = \sup_{\rho \in \mathbf{F}_\nu} \int_0^\infty [x^T Q x + u^T R u] d\tau \leq x_0^T X^{-1}(\rho_0) x_0 \quad (34)$$

While this approach yields a stabilising controller with guaranteed performance bounds, it is potentially conservative because: (i) it uses a quadratic parameter dependent Lyapunov function $x^T X^{-1}(\rho)x$; and (ii) it allows for all possible combinations of the parameters and their derivatives. In the sequel, we indicate how performance can be improved by exploiting the Receding Horizon ideas presented in this chapter.

Lemma 3: Consider the case of LPV dynamics. If the terminal penalty Ψ in Algorithm 3 is chosen as $\Psi(x, \rho) = x^T X^{-1}(\rho)x$, where X denotes any feasible solution to the set of affine matrix inequalities (AMIs) (32), then the following holds:

1. The resulting control law $u_\Psi(x, \rho)$ is globally stabilising for any choice of the horizon T.
2. For any admissible parameter trajectory ρ we have that

$$J_\Psi(x_0, \rho) := \int_0^\infty [x^T Q x + u^T R u] d\tau \leq x^T(0) X^{-1}(\rho_0) x(0) \quad (35)$$

   i.e. the proposed control law is guaranteed to perform no worse than the AMI based control law (33), in the sense that both have the same worst-case upper bound.

Proof: To establish the first claim, note that from the Euler-Lagrange conditions for optimality it can be easily shown that in this case the constraint (18) is redundant, since it is satisfied by the control that optimises the performance index (17). Stability follows now from Theorem 2.

To complete the proof, let x_{AMI} and u_{AMI} denote the trajectory and control corresponding to the parameter trajectory ρ, obtained when using the AMI-based (33) and proposed control law, respectively.

Then, from the definition of u_Ψ it follows that

$$\begin{aligned}
&\int_0^\infty [x_\Psi^T Q x_\Psi + u_\Psi^T R u_\Psi]dt \\
&\quad \leq \int_0^T [x_\Psi^T Q x_\Psi + u_\Psi^T R u_\Psi]dt + x_\Psi^T(T)X^{-1}[\rho(T)]x_\Psi(T) \\
&\quad \leq \int_0^T [x_{AMI}^T Q x_{AMI} + u_{AMI}^T R u_{AMI}]dt + x_{AMI}^T(T)X^{-1}[\rho(T)]x_{AMI}(T).
\end{aligned} \tag{36}$$

From Schur complements, the set of inequalities (32) is equivalent to

$$-\sum_{i=1}^{n_\rho} \nu_i \frac{\partial X}{\partial \rho} + A[\rho(t)]X(\rho) + X(\rho)A^T[\rho(t)] + X(\rho)QX(\rho) - BR^{-1}B^T \leq 0 \tag{37}$$

for all $\underline{\nu}_i \leq \nu_i \leq \bar{\nu}_I$ or, equivalently,

$$-\sum_{i=1}^{n_\rho} \nu_i \frac{\partial X}{\partial \rho} + A_{CL}X + XA_{CL}^T + XQX - BR^{-1}B^T \leq 0 \tag{38}$$

where $A_{CL} = A[\rho(t)] - \frac{1}{2}B[\rho(t)]R^{-1}B^T[\rho(t)]X(\rho)$ denotes the closed-loop dynamic matrix. Pre- and post-multiplying this last equation by $x_{AMI}^T X^{-1}$ and $X^{-1}x_{AMI}$ yields (after some algebra):

$$\frac{d}{dt}\left(x_{AMI}^T X^{-1} x_{AMI}\right) + x_{AMI}^T Q x_{AMI} + u_{AMI}^T R u_{AMI} \leq 0 \tag{39}$$

Finally, integrating this last inequality yields

$$\begin{aligned}
&\int_0^T [x_{AMI}^T Q x_{AMI} + u_{AMI}^T R u_{AMI}]dt + x_{AMI}^T(T)X^{-1}[\rho(T)]x_{AMI}(T) \\
&\quad \leq x^T(0)X^{-1}x(0)
\end{aligned} \tag{40}$$

which, combined with (36), yields the desired result.

4.6 Connections with other approaches

In this section we briefly explore the connections between the proposed controller and some related approaches proposed in the past.

The basic idea of Algorithm 1, namely (i) to convert the infinite dimensional quadratic regulation problem to a finite dimensional optimisation by using a penalty function to estimate the cost-to-go, and (ii) to use explicit optimisation to drive the system to an invariant neighbourhood of the origin and then switch controllers was proposed in References 16 and 17 for the case of constrained LTI systems (see also References 24 and 25 for later work along these lines).

The combination of dual mode control and receding horizon for stabilising nonlinear systems was proposed in Reference 13. However, while this approach has the ability to robustly stabilize nonlinear plant, it does not address the issue of performance. Indeed, the performance index used there (and hence the control action) does not approach the optimal unless $T \to \infty$. From a practical stand-point, this implies that, in order to achieve acceptable performance, the on-line optimisation must be performed over a large horizon, which may not be feasible for moderately large plants or plants with fast time constants.

The extension of the techniques proposed in Reference 16 to nonlinear plants was pursued in Reference 14. Here the infinite horizon cost is approximated by an expression of the form (11), where the terminal penalty function is obtained assuming that a stabilising control law of the form $u = -Kx$ is available and will be used after the horizon T. Alternatively, the penalty function can be obtained by solving the Ricatti equation corresponding to the linearisation of the dynamics around the equilibrium point $x = 0$ (see also Reference 26). Thus, this approach can be viewed as a special case of Algorithm 1 for a particular choice of (local) CLF.

The combination of Receding Horizon and CLF techniques has been proposed independently in References 15, 19 and 27. A difference between these approaches is that the latter does not incorporate a terminal penalty in the performance index. Rather, Reference 15 optimises a performance index of the form

$$J = \int_t^{t+T} [x^T Qx + u^T Ru] d\tau \tag{41}$$

subject to

$$L_f V(x) + L_g V(x) u \leq -\varepsilon\sigma(x);\ V[x(t+T)] \leq V[x_\sigma(t+T)] \tag{42}$$

where $V(\cdot)$ is a CLF and x_σ is the trajectory corresponding to the pointwise minimum norm control that renders $\dot{V} < 0$. This approach is guaranteed to stabilise the system for any horizon T such that the constraints (42) are feasible. However, it may do so at the expense of performance. Note that, contrary to (11), (41) does not approximate the original cost, even if T is taken large or V coincides with the actual value function for the problem. Thus, the corresponding control action, while stabilising, is not necessarily close to the optimal.

Finally, we close this section by showing that, in the limit as $T \to 0$, the control law obtained from Algorithm 3 reduces to the inverse optimal controller proposed by Freeman and Kokotovic [18]. To this effect, note that if $T \to 0$ in (17) then u_Ψ is given by the solution to the following optimisation problem:

$$u_\Psi = \arg\min_u \{u^T R u + L_g \Psi u\} \tag{43}$$

subject to

$$\Psi_0 + \Psi_1 u = 0 \tag{44}$$

where:

$$\begin{aligned} \Psi_0 &= L_f \Psi[x(t)] + x^T(t) Q x(t) + \sup_{\underline{\nu}_i \le \dot{\rho} \le \bar{\nu}_i} \sum_{i=1}^{n_\rho} \frac{\partial \Psi}{\partial \rho_i} \dot{\rho}_i + \alpha[x(t)] \\ \Psi_1 &= L_g \Psi[x(t)] \end{aligned} \tag{45}$$

and where $-\alpha(x)$ is the desired negativity margin [18]. The solution to this optimisation problem is given by

$$u_\Psi = -\frac{\Psi_0 R^{-1} \Psi_1^T}{\Psi_1 R^{-1} \Psi_1^T} \tag{46}$$

This is precisely the inverse optimal controller obtained in Reference 18.

4.7 Incorporating constraints

Next, we briefly discuss how to incorporate constraints into the formalism, proceeding in the spirit of References 9 and 16.

Assume, for instance, that the control action is constrained to belong to a given compact, convex, possibly parameter dependent set $u \in \mathbf{U}(\rho)$. Let V_u denote the unconstrained value function for the problem in a region S, and consider the set $\mathbf{U}_V = \left\{x : u = -\frac{1}{2} g^T \frac{\partial V_u}{\partial x} \in \mathbf{U}(\rho) \text{ for all } \rho \in \mathbf{P}\right\}$, i.e. the set of states where the unconstrained control law satisfies the control constraints. Finally, denote by S_V the largest invariant set contained in U_V. Then, Algorithm 1 can be applied to the problem by simply modifying the explicit optimisation (11) to take the constraints into account and using the set S_V as S_v. Clearly, this modified control law has the same properties as Algorithm 1, i.e. if the optimisation is feasible, then it stabilises the system and yields the optimal control action. Similarly, the use of a fixed optimisation horizon T and an approximation $\Psi(x, \rho)$ can be taken into account by selecting Ψ to be a constrained control Lyapunov function in the sense that

$$\min_{u \in \cap \mathbf{U}(\rho)} \sup_{\rho \in \mathbf{P}, \underline{\nu}_i \le \nu \le \bar{\nu}_i} \left\{ \frac{\partial \Psi}{\partial x}[f(x,\rho) + g(x,\rho)u] + \sum_i \frac{\partial \Psi}{\partial \rho_i} \nu_i \right\} \le -\underline{\sigma} \|x\|^2 < 0 \tag{47}$$

$$\forall x \in S_v$$

and modifying the algorithm so that the minimisation is taken over $u \in \mathbf{U}(\rho)$ rather than over $\mathbf{R}^{n_u}$. Clearly, this algorithm stabilises the system in the region where this can be accomplished with bounded controls. Finally, state constraints can also be incorporated into the formalism by rendering suitable sets invariant, proceeding as in Reference 9.

4.8 Illustrative examples

This section illustrates our results with several examples. The first one is a simple academic example that can be solved analytically. Thus it can be used to explicitly analyse the source of performance degradation when using feedback linearisation or the SDRE method, and to show the advantages of the proposed approach. The second example, a realistic problem arising in the context of control of thrust vectored aircraft, was used for benchmarking several nonlinear design methods in Reference 5. Finally, the third example illustrates the advantages of the proposed method for the case of LPV systems.

Example 1: Consider the following regulation problem:

$$\min_u \left\{ J + \int_0^\infty [x_2^2 + u^2] dt \right\} \tag{48}$$

subject to

$$\begin{aligned} \dot{x}_1 &= x_2 \\ \dot{x}_2 &= -x_1 e^{x_1} + \frac{1}{2} x_2^2 + e^{x_1} u \end{aligned} \tag{49}$$

It can be shown that the optimal control law is given by

$$u_{opt} = -x_2 \tag{50}$$

with optimal storage function

$$V(x) = x_1^2 + x_2^2 e^{-x_1} \tag{51}$$

A feedback linearisation design selected so that the closed-loop system has the same storage function as $\|x\| \to 0$ yields the following controller and Lyapunov function:

$$\begin{aligned} u_{FL} &= (1 - e^{-x_1})x_1 - x_2 e^{-x_1}(1 + 0.5x_2) \\ V_{FL} &= x_1^2 + x_2^2 \end{aligned} \tag{52}$$

Note that $u_{FL} \simeq u_{opt}$ only for small values of x_1 and x_2. Consider now the following state-dependent coefficient (SDC) parameterisation:

$$A = \begin{bmatrix} 0 & 1 \\ -e^{x_1} & \frac{1}{2}x_2 \end{bmatrix}; \quad B = \begin{bmatrix} 0 \\ e^{x_1} \end{bmatrix} \tag{53}$$

It can be shown that the solution to the corresponding SDRE is given by

$$P(x) = \begin{bmatrix} e^{x_1} & 0 \\ 0 & 1 \end{bmatrix} p(x) \tag{54}$$

where

$$p(x) = \left[\frac{x_2}{2e^{x_1}} + \sqrt{1 + \left(\frac{x_2}{2e^{x_1}} \right)^2} \right] e^{-x_1} \tag{55}$$

with associated control action

$$u_{SDRE} = -x_2 \left[\left(\frac{x_2}{2e^{x_1}} \right) + \sqrt{1 + \left(\frac{x_2}{2e^{x_1}} \right)^2} \right] \tag{56}$$

Finally it can also be shown that

$$x^T P(x) x = p(x) e^{x_1} V(x) \tag{57}$$

Thus $x^T P(x) x$ gives a good estimate of $V(x)$ and $u_{SDRE} \simeq u_{opt}$ only when $x_2/2e^{x_1} \ll 1$.

Table 4.2 shows the different costs starting from the initial condition $x(0) = [-2 \quad 2]^T$ for several controllers, with the corresponding trajectories

Table 4.2 Comparison of different methods for Example 1

Method	*Cost*
Optimal	33.56
FL	95.11
SDRE	143.00

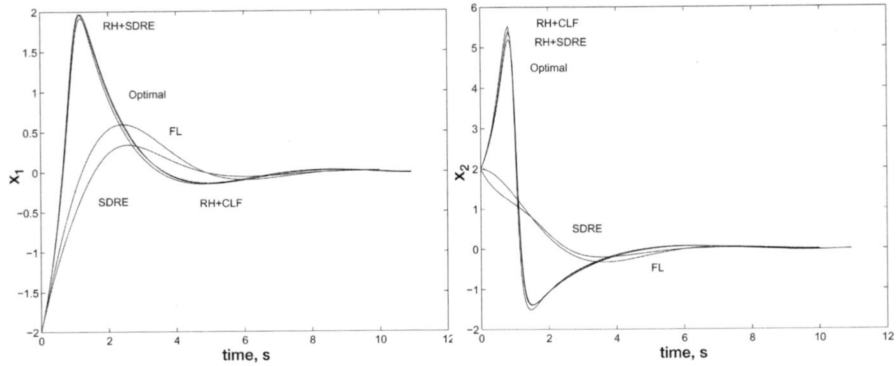

Figure 4.2 State trajectories for Example 1

shown in Figures 4.2 and 4.3. The last two entries of the table correspond to the proposed controller using a horizon $T = 1$ s and as estimates of the value function $\Psi = V_{FL}$ and $\Psi = x^T P(x)x$, respectively. Note that in this case performance of both controllers is virtually identical to optimal.

Example 2: Consider again the simplified model of the thrust vectored aircraft used in the introduction. Table 4.3 shows the cost corresponding to the initial condition $\xi(0) = [0 \quad 0 \quad 0 \quad 12.5 \quad 0 \quad 0]^T$, obtained using different controllers. The two lowest entries correspond to the proposed method using $T = 1$ s and $T_s = 0.5$ s and terminal penalties derived from Jacobian linearisation and the SDRE methods, respectively. Note that the latter virtually achieves optimal performance, while the former is only 2 per cent suboptimal. This behaviour can be explained by

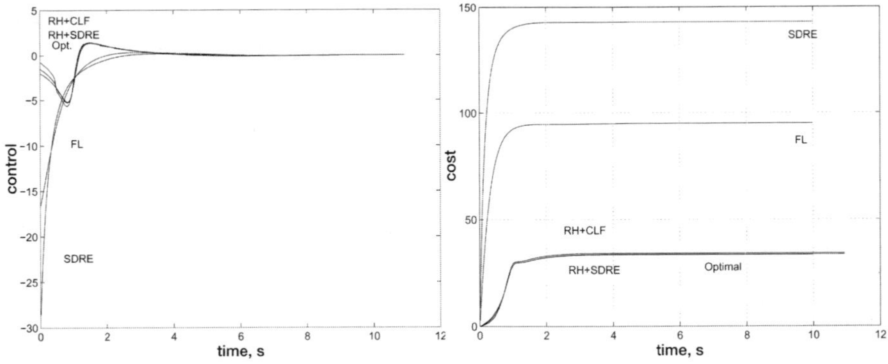

Figure 4.3 Control effort and cost for Example 1

Table 4.3 Comparison of different methods for Example 2

Method	*Cost*
Exact	1115
LQR [5]	1.1×10^5
CLF [5]	2.53×10^4
LPV [5]	1833
SDRE	1640
RH + JL	1142 ($T = 1$)
	1321 ($T = 0.4$)
RH + SDRE	1117 ($T = 1$)
	1310 ($T = 0.4$)

looking at Figure 4.4, which shows the different portions of the cost as a function of the horizon, starting from the initial condition $\xi(0)$. These plots show that, while $\Psi(x) = x^T P(x)x$ gives initially a very poor estimate of the cost-to-go, the combination of $\Psi(x)$ and the explicit integral in (17) give a very good estimate if T is chosen ≥ 1 s. It is worth mentioning that a conventional receding horizon controller (i.e. one obtained by setting $\Psi \equiv 0$ in (17)) with the same choice of horizon and sampling time fails to stabilize the system.

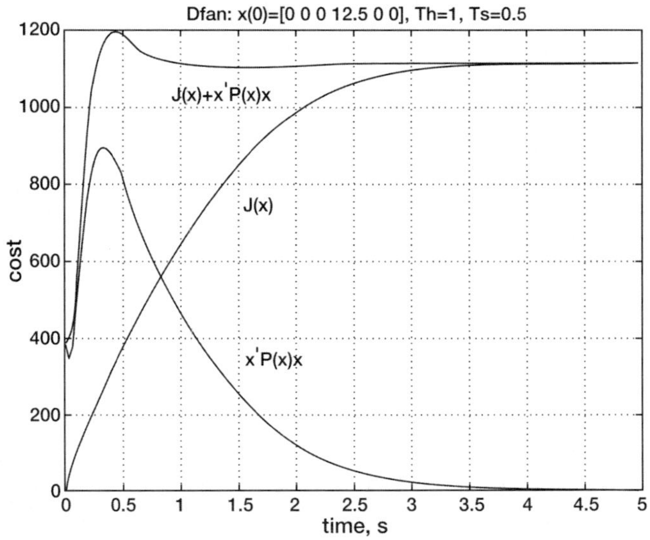

Figure 4.4 The terms of the cost as functions of the horizon in Example 2

Example 3: Consider an LPV system with the following state space realisation:

$$A = \begin{bmatrix} 0 & 1 \\ 1 & 0.5\rho - 1.5 \end{bmatrix}, \quad B_2 = \begin{bmatrix} 0 \\ 1 \end{bmatrix}$$
$$C_1 = \sqrt{2}\begin{bmatrix} 1 & -1 \\ 0 & 0 \end{bmatrix}, \quad D_{12} = \begin{bmatrix} 0 \\ 1 \end{bmatrix} \tag{58}$$
$$P = \{\rho : 0 \leq \rho \leq 1\}; \quad \underline{\nu} = -2, \bar{\nu} = 2$$

It can be easily verified that the following matrix function satisfies the AMIs (32):

$$X(\rho) = X_0 + X_1\rho + X_2\rho^2$$
$$X_0 = \begin{bmatrix} 0.2210 & -0.3505 \\ -0.3505 & 1.1272 \end{bmatrix}, \quad X_1 = \begin{bmatrix} -0.0239 & 0.0924 \\ 0.0924 & -0.3577 \end{bmatrix}$$
$$X_2 = \begin{bmatrix} 0.0243 & -0.0638 \\ -0.0683 & 0.2180 \end{bmatrix} \tag{59}$$

for all $\rho \in \mathbf{F}_\nu$. Figure 4.5 compares the trajectories starting from the initial condition $[0 \quad 2]^T$ for the AMI-based $(x_{1lpv}, x_{2lpv}, u_{lpv})$ and the proposed controller,

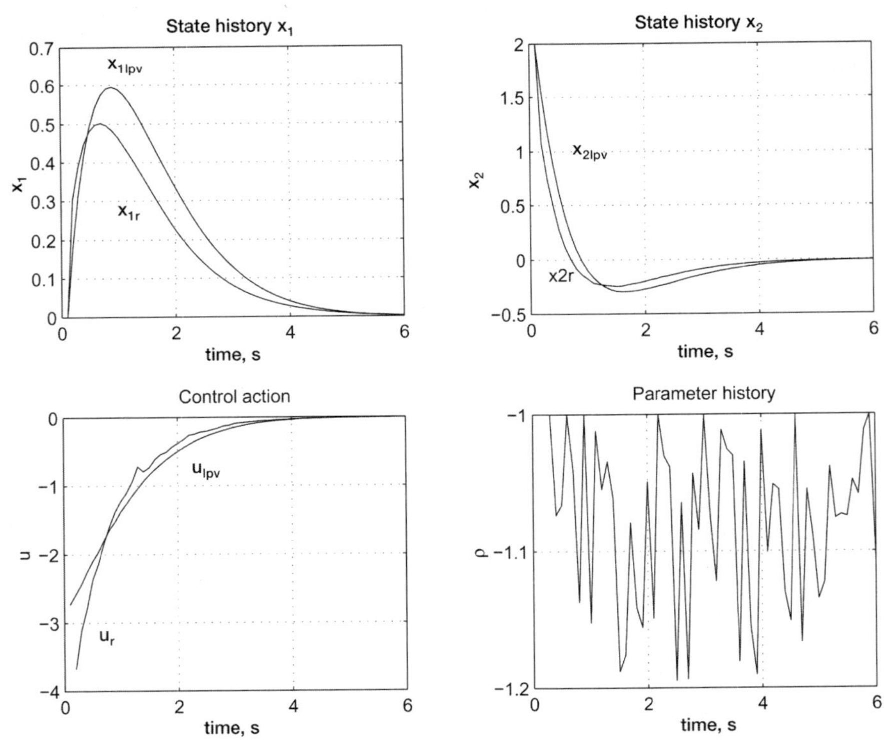

Figure 4.5 State, control and (normalised) parameter trajectories for Example 3

respectively (in this figure the parameter was normalised to $\rho_n = 0.5\rho - 1.5$). The latter was implemented using $T = 2$ as horizon and $\Psi = x^T X^{-1}(\rho)x$. For the specific parameter history shown there, the receding horizon controller yields $J = 6.91$ versus $J = 8.30$ for the control law (33), a performance improvement of roughly 20 per cent. Similar results were obtained for other initial conditions and parameter trajectories.

4.9 Conclusions

In contrast with the case of linear plants, tools for simultaneously addressing performance and stability of nonlinear systems have emerged relatively recently. Recent counterexamples [5] illustrated the fact that, while several commonly used techniques can successfully stabilise nonlinear systems, the resulting closed-loop performance varies widely. Moreover, these performance differences are problem dependent, with performance of a given method ranging from (near) optimal to very poor.

In this chapter we have proposed a new suboptimal regulator for control affine parameter dependent nonlinear systems, based upon the combination of receding horizon and control Lyapunov functions techniques. The main result of the chapter shows that, under certain relatively mild conditions, essentially equivalent to the existence of a control Lyapunov function, this regulator renders the origin a globally asymptotically stable equilibrium point. Additional results show that some readily available choices for the CLF $\Psi(x, \rho)$ render the proposed controller near optimal for some cases of practical interest. These results were illustrated with a number of examples where the proposed controller out-performed several other commonly used techniques.

An issue that was not addressed here is that of the computational complexity associated with solving the nonlinear optimisation problem (17). Following Reference 5 this complexity could be reduced by exploiting differential flatness to perform the optimisation in flat space. Additional research being pursued includes the extension of the framework to the output feedback case and to handle model uncertainty.

4.10 Acknowledgments

This work was supported in part by NSF grants ECS-9625920 and ECS-9907051 and AFOSR under grant F49620-00-1-0020.

4.11 References

1 KRSTIC, M., KANELLAKOPOULOS, I., and KOKOTOVIC, P.: ‘Nonlinear and adaptive control design’ (John Wiley, New York, 1995)

2 ARTSTEIN, Z.: 'Stabilization with relaxed controls', *Nonlinear Analysis*, 1983, **7** (11), pp. 1163–73

3 SONTAG, E.D.: 'A universal construction of Artstein's Theorem on nonlinear stabilization', *Syst. Control Lett.*, 1989, **13** (2), pp. 117–23

4 VAN NIEUWSTADT, M.J., and MURRAY, R.M.: 'Real time trajectory generation for differentially flat systems', *Int. J. Robust Nonlinear Control*, 1998, **8** (11), pp. 995–1020

5 DOYLE, J.C. *et al.*: 'Nonlinear control: comparisons and case studies'. Workshop #7, 1997 ACC, Albuquerque, NM, June 1997

6 BALL, J.A., HELTON, J.W., and WALKER, M.L.: 'H_∞ control for nonlinear systems with output feedback', *IEEE Trans. Autom. Control*, 1993, **38** (4), pp. 546–59

7 ISIDORI, A., and WANG, K.: 'H_∞ control via measurement feedback for general nonlinear systems', *IEEE Trans. Autom. Control*, 1992, **37** (9), pp. 1283–93

8 LU, W.M., and PACKARD, A.: 'Asymptotic rejection of persistent L_∞ bounded disturbances for nonlinear systems'. Proceedings of 35th IEEE CDC, Kobe, Japan, December 1996, pp. 2401–6

9 SZNAIER, M., and DAMBORG, M.: 'Heuristically enhanced feedback control of constrained discrete time linear systems', *Automatica*, 1990, **26** (3), pp. 521–32

10 ROSSITER, J.A., and KOUVARITAKIS, B.: 'Constrained generalised predictive control', *IEE Proc. D*, 1993, **140**, pp. 233–54

11 RAWLINGS, J.B., and MUSKE, K.R.: 'Stability of constrained receding horizon control', *IEEE Trans. Autom. Control*, 1993, **38** (10), pp. 1512–6

12 MAYNE, D.Q., and MICHALSKA, H.: 'Receding horizon control of nonlinear systems', *IEEE Trans. Autom. Control*, 1990, **35** (7), pp. 814–24

13 MICHALSKA, H., and MAYNE, D.Q.: 'Robust receding horizon control of constrained nonlinear systems', *IEEE Trans. Autom. Control*, 1993, **38** (11), pp. 1623–33

14 MAGNI, L.: 'Nonlinear receding horizon control: theory and application'. PhD Thesis, University of Pavia, 1997

15 PRIMBS, J.A., NEVISTIC, V., and DOYLE, J.C.: 'On receding horizon extensions and control Lyapunov functions'. Proceedings of 1998 ACC, June 1998, pp. 3276–80

16 SZNAIER, M., and DAMBORG, M.J.: 'Suboptimal control of linear systems with state and control inequality constraints'. Proceedings of 26th IEEE CDC, Los Angeles, CA, December 1987, pp. 761–2

17 SZNAIER, M.: 'Suboptimal feedback control of constrained linear systems'. PhD Dissertation, University of Washington, Seattle, WA, 1989

18 FREEMAN, R.A., and KOKOTOVIC, P.V.: 'Inverse optimality in robust stabilization', *SIAM J. Control Optim.*, 1996, **34**, pp. 1365–91

19 SZNAIER, M.: 'Suboptimal control of nonlinear systems via receding horizon state dependent Riccati equations'. Final Report, AFOSR Summer Faculty Reseach Program, Wright Laboratory, August 1997

20 FREEMAN, R.A., and PRIMBS, J.A.: 'Control Lyapunov functions: new ideas from an old source'. Proceedings of 35th IEEE CDC, Kobe, Japan, December 1996, pp. 3926–31

21 FREEMAN, R.A., and KOKOTOVIC, P.V.: 'Robust nonlinear control design: state space and Lyapunov techniques' (Birkhauser, Boston, 1996)

22 WU, F., YANG, X.H., PACKARD, A., and BECKER, G.: 'Induced L_∞ norm control for LPV systems with bounded parameter variation rates', *Int. J. Robust Nonlinear Control*, 1996, **6** (9/10), pp. 983–98
23 BALAS, G., *et al.* Lecture Notes, Workshop on Theory and Application of Linear Parameter Varying Techniques, 1997 ACC
24 CHMIELEWSKI, D., and MANOUSIOUTHAKIS, V.: 'On constrained infinite-time linear quadratic optimal control', *Syst. Control Lett.*, 1996, **29** (3), pp. 121–9
25 SCOKAERT, P., and RAWLINGS, J.: 'Constrained linear quadratic regulation', *IEEE Trans. Autom. Control*, 1998, **43** (8), pp. 1163–9
26 CHEN, H., SCHERER, C., and ALLGÖWER, F.: 'A game theoretical approach to nonlinear robust receding horizon control of constrained systems'. Proceedings of 1997 ACC, June 1997, pp. 3073–7
27 SZNAIER, M., CLOUTIER, J., HULL, R., JACQUES, D., and MRACEK, C.: 'A receding horizon state dependent Riccati equation approach to suboptimal regulation of nonlinear systems'. Proceedings of 37th IEEE CDC, December 1998, pp. 1792–7
28 CLOUTIER, J.R., D'SOUZA, C.N., and MRACEK, C.P.: 'Nonlinear regulation and nonlinear H_∞ control via the state-dependent Riccati equation technique: Part 1, theory; Part 2, examples'. Proceedings of the International Conference on Nonlinear Problems in Aviation and Aerospace, Daytona Beach, FL, May 1996, pp. 117–42
29 MRACEK, C.P., and CLOUTIER, J.R.: 'Control design for the nonlinear benchmark problem via the state-dependent Riccati equation method', *Int. J. Robust Nonlinear Control*, 1998, **8**, pp. 401–33
30 JACOBSON, D.H.: 'Extensions of linear quadratic control, optimization and matrix theory' (Academic Press, London, 1977)
31 KHALIL, H.K.: 'Nonlinear systems' (Prentice Hall, 1996, 2nd edn)

4.12 Appendix: SDRE approach to nonlinear regulation

In this section we briefly cover the details of the SDRE approach developed by Cloutier and coworkers [28, 29]. The main idea of the method is to recast the nonlinear system (5) into a state dependent coefficient (SCD) linear-like form:

$$\dot{x} = A(x)x + B(x)u \tag{60}$$

and to solve pointwise along the trajectory the corresponding algebraic Riccati equation:

$$A^T(x)P(x) + P(x)A(x) - P(x)B(x)R^{-1}(x)B^T(x)P(x) + Q(x) = 0 \tag{61}$$

The suboptimal control law is given by $u_{SDRE} = -\frac{1}{2}R^{-1}(x)B^T(x)P(x)x$, where $P(x)$ is the positive definite (pointwise stabilising) solution of (61). In the sequel we briefly review the properties of this control law. The corresponding proofs can be found in the appropriate references.

Lemma 4 ([28, 29]): Assume that $Q(x) = C^T(x)C(x)$ and that there exists a neighbourhood Ω of the origin where the pairs $\{A(x), B(x)\}$ and $\{A(x), C(x)\}$ are pointwise stabilisable and detectable, respectively, and all the matrix functions involved are C^1. Then the control law u_{SDRE} renders the origin a locally asymptotically stable equilibrium point of the closed-loop system.

Lemma 5 ([28, 29]): The SDRE control law and its associated state and co-state trajectories satisfy the following necessary condition for optimality: $\frac{\partial H}{\partial u} = 0$ where $H = x^T Q(x)x + u^T R(x)u + \lambda^T[f(x) + B(x)u]$ is the Hamiltonian of the system and where λ denotes the co-states.

Lemma 6 ([6, 19]): Assume that the parameterisation (60) is stabilisable and all the matrices involved along with their gradients are bounded in a neighbourhood Ω of the origin. Then the SDRE control law and its associated state and co-state trajectories asymptotically satisfy at a quadratic rate (i.e. $\|\dot{\lambda} + \frac{\partial H}{\partial x}\| \to 0$ as $O(\|x\|^2)$ as $x \to 0$) the following necessary condition for optimality: $\dot{\lambda} = -\frac{\partial H}{\partial x}$ in the sense that

$$\left\|\dot{\lambda} + \frac{\partial H}{\partial x}\right\| \leq x^T U x$$

for some matrix $U > 0$ and all $x \in \Omega$.

Lemma 7 ([30]): Let $P(x)$ denote a solution to the SDRE (61). If there exists a positive definite function $V(x)$ such that $\frac{\partial V(x)}{\partial x} = P(x)x$ then u_{SDRE} is the globally optimal control law. (A necessary and sufficient condition for this to hold is that the Jacobian matrix $\frac{\partial}{\partial x}[P(x)]$ is symmetric (see, for instance, Reference 31, Section 3.1). In this case $V(x)$ can be computed as $V(x) = \int_0^x y^T P(y)dy$, where this line integral is independent of the path.)

Chapter 5

Nonlinear model-algorithmic control for multivariable nonminimum-phase processes

Michael Niemiec and Costas Kravaris

Abstract

A general nonlinear discrete-time controller design for multivariable nonminimum-phase systems is presented that utilises a class of synthetic outputs that are statically equivalent to the process outputs and makes the system minimum-phase. A systematic procedure is outlined for the selection of particular synthetic outputs and leads to prescribed transmission zeros. The selected synthetic outputs are then used to construct a model-algorithmic controller, which is guaranteed to induce zero steady-state error between the setpoints and original plant outputs. The proposed control scheme shows excellent performance in a simulation case study for a multivariable nonisothermal CSTR where a series/parallel reaction is taking place.

5.1 Introduction

Processes that exhibit strong nonlinearities dominate the chemical engineering field, creating major challenges in design and control. Since the control algorithm is usually based upon a linear approximation of the process dynamics around an operating steady state, performance can be unacceptably poor in the presence of strong nonlinearities. Processes with strong nonlinearities include polymerisation reactors, high-purity distillation columns, bioprocesses and pH processes. The significance of nonlinearities and the insufficiency of traditional control strategies have shifted academic and industrial efforts to develop new nonlinear control schemes. These schemes include model predictive process control and geometric process control. As a direct result, the area of nonlinear process control has evolved as a structurally solid research area with practical applicability, enabling the solution of important practical problems. One class of nonlinear controllers, which combines the theoretical rigour of geometric control methods with the intuitive appeal of model-predictive control methods, is the class of model-algorithmic controllers (MAC). In its original form, model-algorithmic control is a model-predictive controller, which uses a linear impulse response model to predict the future behaviour of the system. MAC was developed in France in the late 1960s within the chemical process industry. It originally appeared as a heuristic algorithm under the name of Model Heuristic Control in the late 1970s [1], with advancements in the theory occurring in the early 1980s [2, 3]. Although the original development of MAC uses a linear input/output model, a generalised state-space approach for both linear and nonlinear systems has been developed following the conceptual steps of the original methodology [4, 5].

Model-algorithmic control belongs to the family of control strategies that force the controlled outputs to follow a prespecified response in closed loop and therefore generate an inverse operator in an explicit or implicit manner. For inverse-based or trajectory-following control algorithms, special modifications must be made when they are applied to nonminimum-phase systems where the inverse is unstable; this is commonly referred to as nonminimum-phase compensation.

Research in the area of nonminimum-phase compensation is still in the early stages and is mainly confined to continuous-time systems. For linear systems, the system is factored into minimum-phase and nonminimum-phase parts, with the minimum-phase part inverted for the controller design. For nonlinear systems, the decomposition into minimum-phase and nonminimum-phase parts is an extremely difficult problem, even in the SISO case. In the special case of SISO second-order systems, the decomposition problem was solved and ISE-optimal control laws were formulated [6]. Alternative nonminimum-phase compensation approaches have also been developed which include approximate stable/anti-stable factorisation of the zero dynamics, an inner-outer based approximation and a multiple-input approach; these are also applicable to limited classes of nonlinear systems [7, 8].

In a different line of research, a nonminimum-phase compensation structure for SISO nonlinear systems was developed that is based on a synthetic output which makes the system minimum-phase and is statically equivalent to the original output [9, 10]. The synthetic output is calculated on-line and controlled to setpoint via an inverse-based controller, bypassing the difficulty of decomposition and reducing the problem to the construction of an appropriate synthetic output. To this end, an ISE-optimal formulation was proposed that led to analytical results only for a limited class of nonlinear SISO systems. This concept was extended when the search for an ISE-optimal output was relaxed and a statically equivalent output was constructed with prescribed zeros [11]. The present work generalises the methods and results of Niemiec and Kravaris [12, 13] to nonlinear discrete-time multivariable nonminimum-phase systems.

5.2 Preliminaries

Consider a discrete-time multivariable nonlinear system of the form:

$$\begin{aligned} x(k+1) &= \Phi[x(k), u(k)] \\ y(k) &= h[x(k)] \end{aligned} \tag{1}$$

where x denotes the vector of state variables, u denotes the manipulated input vector, and y denotes the controlled output vector. It is assumed that $x \in X \subset \Re^n$, $u \in U \subset \Re^m$ and $\Phi : X \times U \to \Re^n, h : X \to \Re^m$ are analytic functions on $X \times U$ and X, respectively. Moreover, the following standing assumptions will be made:

(A1) $$\text{Rank}\left[\frac{\partial \Phi}{\partial u}(x, u)\right] = m \text{ for all } (x, u) \in X \times U$$

(A2) $$\text{Rank}\left[\frac{\partial h}{\partial x}(x)\right] = m \text{ for all } x \in X$$

The equilibrium set of a system of the form (1) is defined by

$$E = \{x \in X | \exists u \in U \text{ satisfying } \Phi(x, u) = x\}$$

Because of assumption (A1), E is an m-dimensional manifold.

A model of the form (1) can be obtained either from first principles (e.g. in chemical processes, using material and energy balances together with appropriate rate expressions for the physical and chemical phenomena), or using system identification techniques. In the former case, the original model is usually a continuous-time affine system of the form:

$$\begin{aligned} \dot{x} &= f(x) + g(x)u \\ y &= h(x) \end{aligned} \tag{2}$$

which has been subsequently discretised under the zero-order hold assumption using standard numerical methods (e.g. Euler, Taylor Series, Runge-Kutta, etc.).

Two important remarks must be made here relative to systems of the form (1) which are approximate sampled-data representations of systems of the form (2). First, unless Euler's method is used in the derivation of the sampled-data representation, the discrete-time system will have the manipulated input appearing nonlinearly in the right-hand side. Second, in all approximate discretisation methods, the continuous-time system and the time-discretised system have the same equilibrium characteristics and therefore the same equilibrium manifold. Consequently, the equilibrium manifold may be calculated from the original continuous-time system as

$$E = \{x \in X | \exists u \in U \text{ satisfying } f(x) + g(x)u = 0\}$$

which is likely to be easier because of the affine dependence of the defining equation on u.

5.2.1 Relative order

The notion of relative order [14] will play an instrumental role in the development of the control algorithm and therefore it needs to be reviewed here.

The *i*th output y_i of a system of the form (1) is said to have relative order 1 if

$$\frac{\partial}{\partial u} h_i[\Phi(x,u)] \not\equiv [0 \ldots 0]$$

Otherwise, the function composition $h_i[\Phi(x,u)]$ is independent of u and therefore it defines a function of x only:

$$h_i^1(x) = h_i[\Phi(x,u)]$$

If it so happens that

$$\frac{\partial}{\partial u} h_i^1[\Phi(x,u)] \not\equiv [0 \ldots 0]$$

then the output y_i is said to have relative order 2. Otherwise, the function composition $h_i^1[\Phi(x,u)]$ is independent of u and therefore it defines a function of x only:

$$h_i^2(x) = h_i^1[\Phi(x,u)]$$

This procedure continues, defining functions

$$h_i^l(x) = h_i^{l-1}[\Phi(x,u)]$$

up until, for some integer r_i,

$$\frac{\partial}{\partial u} h_i^{r_i-1}[\Phi(x,u)] \not\equiv [0 \ldots 0] \tag{3}$$

in which case the output y_i is said to have relative order r_i. If no such integer exists, the output y_i is said to have infinite relative order.

For a system of the form (1) with finite relative order r_i for the output y_i, the following relations hold:

$$\begin{aligned} y_i(k+1) &= h_i^1[x(k)] \\ y_i(k+2) &= h_i^2[x(k)] \\ &\vdots \\ y_i(k+r_i-1) &= h_i^{r_i-1}[x(k)] \\ y_i(k+r_i) &= h_i^{r_i-1}[\Phi[x(k),u(k)]] \end{aligned} \tag{4}$$

If the process output y_i does not have a finite relative order $(r_i = \infty)$, the manipulated input vector does not affect the output y_i. In a well-formulated control problem, every output y_i must possess a finite relative order r_i.

The characteristic matrix of a system of the form (1) with finite relative orders r_i is defined as

$$C(x,u) \triangleq \begin{bmatrix} \frac{\partial}{\partial u} h_1^{r_1-1}[\Phi(x,u)] \\ \vdots \\ \frac{\partial}{\partial u} h_m^{r_m-1}[\Phi(x,u)] \end{bmatrix} \tag{5}$$

Throughout this work, it will be assumed that

(A3) $\det C(x,u) \neq 0$ for all $(x,u) \in X \times U$

Because of the above assumption, the nonlinear algebraic equations

$$h_i^{r_i-1}[\Phi(x,u)] = \tilde{y}_i, \quad i = 1, \ldots, m \tag{6}$$

are locally solvable for the manipulated input vector u. The implicit function will be denoted by

$$u = \Psi_o(x, \tilde{y}) \tag{7}$$

5.2.2 Zero dynamics and minimum-phase behaviour

The notion of relative order motivates the decomposition of the system into two subsystems in series: (i) a delay-free subsystem and (ii) a pure delay subsystem:

$$\left.\begin{aligned} x(k+1) &= \Phi[x(k), u(k)] \\ \tilde{y}_i(k) &= h_i^{r_i-1}\{\Phi[x(k), u(k)]\}, \quad i = 1, \ldots, m \end{aligned}\right\} \text{delay-free}$$
$$y_i(k) = \tilde{y}_i(k - r_i), \qquad i = 1, \ldots, m\} \text{ pure delay} \tag{8}$$

This decomposition is a generalisation of the factorisation of linear discrete-time systems into an invertible part and a pure delay. The inverse of the delay-free subsystem can be represented as:

$$\begin{aligned} x(k+1) &= \Phi\{x(k), \Psi_o[x(k), \tilde{y}(k)]\} \\ u(k) &= \Psi_o[x(k), \tilde{y}(k)] \end{aligned} \tag{9}$$

If the system (1) is linear, the inverse system [9] has $n - \sum_{i=1}^{m} r_i$ poles at the transmission zeros of (1) and $\sum_{i=1}^{m} r_i$ poles at the origin. For the inverse system to be stable, the finite transmission zeros must be inside the unit circle. Accordingly, stability of the nonlinear inverse system can be defined.

Definition 1: *Given a discrete-time nonlinear system of the form (1), its delay-free part is called minimum-phase if the dynamics*

$$x(k+1) = \Phi\{x(k), \Psi_o[x(k), \tilde{y}(k)]\} \tag{10}$$

is locally asymptotically stable. Otherwise, the system will be called nonminimum-phase.

The local asymptotic stability of (10) can be checked via Lyapunov's first method by calculating the eigenvalues of the Jacobian evaluated at an equilibrium point. Utilising the implicit function theorem, the Jacobian of the system is

$$J(x,u) = \left[\frac{\partial\Phi(x,u)}{\partial x}\right] - \left[\frac{\partial\Phi(x,u)}{\partial u}\right][C(x,u)]^{-1}\begin{bmatrix}\frac{\partial}{\partial x}h_1^{r_1-1}[\Phi(x,u)] \\ \vdots \\ \frac{\partial}{\partial x}h_m^{r_m-1}[\Phi(x,u)]\end{bmatrix} \tag{11}$$

and is evaluated at a reference equilibrium point. If all the eigenvalues of the Jacobian are in the interior of the unit circle, the dynamics of (10) will be locally asymptotically stable around the reference equilibrium point.

Definition 2: *Given a discrete-time nonlinear system of the form (1), its delay-free part is called locally hyperbolically minimum-phase around a reference equilibrium point if all the eigenvalues of the Jacobian matrix, evaluated at that point, are in the interior of the unit circle.*

It is important to note that if a system of the form (1) is hyperbolically minimum-phase, then it will also be minimum-phase. However, the converse of the statement may not hold. The dynamics of (10) can have some eigenvalues on the unit circle and still be asymptotically stable.

5.3 Brief review of nonlinear model-algorithmic control

For a process modelled by a system of the form (1), the future changes of the ith output can be predicted according to:

$$\begin{aligned}
y_{iM}(k+1) - y_{iM}(k) &= h_i^1[x_M(k)] - h_i[x_M(k)] \\
y_{iM}(k+2) - y_{iM}(k) &= h_i^2[x_M(k)] - h_i[x_M(k)] \\
&\vdots \\
y_{iM}(k+r_i-1) - y_{iM}(k) &= h_i^{r_i-1}[x_M(k)] - h_i[x_M(k)] \\
y_{iM}(k+r_i) - y_{iM}(k) &= h_i^{r_i-1}\{\Phi[x_M(k), u(k)]\} - h_i[x_M(k)]
\end{aligned} \tag{12}$$

where r_i is the relative order of the output and the subscript M denotes variables from the on-line simulation of the process model. The predicted changes of the output can be added to the actual output measurement y_i to obtain 'closed-loop' predictions of the future output values:

$$\begin{aligned} \hat{y}_i(k+1) &= y_i(k) + h_i^1[x_M(k)] - h_i[x_M(k)] \\ \hat{y}_i(k+2) &= y_i(k) + h_i^2[x_M(k)] - h_i[x_M(k)] \\ &\vdots \\ \hat{y}_i(k+r_i-1) &= y_i(k) + h_i^{r_i-1}[x_M(k)] - h_i[x_M(k)] \\ \hat{y}_i(k+r_i) &= y_i(k) + h_i^{r_i-1}\{\Phi[x_M(k), u(k)]\} - h_i[x_M(k)] \end{aligned} \tag{13}$$

By using the difference of the predicted output functions, any constant errors associated with the modelling of the process will be eliminated.

The response of each process output is requested to follow a pre-assigned reference trajectory:

$$\hat{y}_i(k+r_i) = (1-\alpha_i)y_{isp} + \alpha_i\hat{y}_i(k+r_i-1) \tag{14}$$

where the subscript *sp* denotes the output setpoint and α_i is a tunable scalar parameter such that $0 \leq \alpha_i < 1$. A parameter of $\alpha_i = 0$ gives rise to a dead-beat controller which requests that the output y_i reach the setpoint in r_i sampling periods. This kind of response has poor robustness properties and is often unreasonable due to constraints on the process inputs.

Combining (13) and (14), so that the 'closed-loop' output prediction matches the reference trajectory, gives a set of nonlinear algebraic equations which depend on the manipulated input vector:

$$\begin{aligned} h_1^{r_1-1}\{\Phi[x_M(k), u(k)]\} &= (1-\alpha_1)\{y_{1sp} - y_1(k) + h_1[x_M(k)]\} + \alpha_1 h_1^{r_1-1}[x_M(k)] \\ &\vdots \\ h_m^{r_m-1}\{\Phi[x_M(k), u(k)]\} &= (1-\alpha_m)\{y_{msp} - y_m(k) + h_m[x_M(k)]\} + \alpha_m h_m^{r_m-1}[x_M(k)] \end{aligned} \tag{15}$$

Because the system has a nonsingular characteristic matrix by assumption (A3), the above nonlinear algebraic equations are locally solvable for $u(k)$ via the implicit function theorem. The corresponding implicit function defined as the solution to (15) is denoted by

$$u(k) = \Psi\{x_M(k), y_{sp} - y(k) + h[x_M(k)]\} \tag{16}$$

With the model states x_M obtained by simulating $x_M(k+1) = \Phi[x_M(k), u(k)]$ on-line, the resulting control law can be expressed as:

$$\begin{aligned} x_M(k+1) &= \Phi(x_M(k), \Psi\{x_M(k), y_{sp} - y(k) + h[x_M(k)]\}) \\ u(k) &= \Psi\{x_M(k), y_{sp} - y(k) + h[x_M(k)]\} \end{aligned} \tag{17}$$

Under this controller, the closed-loop system is given by:

$$\begin{aligned} x_M(k+1) &= \Phi(x_M(k), \Psi\{x_M(k), y_{sp} - y(k) + h[x_M(k)]\}) \\ x(k+1) &= \Phi(x(k), \Psi\{x_M(k), y_{sp} - y(k) + h[x_M(k)]\}) \\ y(k) &= h[x(k)] \end{aligned} \tag{18}$$

The structure of the closed-loop system is depicted in Figure 5.1. It was shown in Soroush and Kravaris [5] that the closed-loop system of (18) will be locally asymptotically stable under the following conditions:

(i) The delay-free part of the system of (1) is locally hyperbolically minimum-phase.
(ii) The open-loop dynamics $x(k+1) = \Phi[x(k), u(k)]$ is locally hyperbolically stable.
(iii) The reference trajectories are stable $(0 \leq \alpha_i < 1$ for $i = 1, \ldots, m)$.

Out of the above conditions, condition (iii) does not pose any restriction on the algorithm; any meaningful reference trajectory will have to be stable. However, condition (i) (minimum-phase behaviour) and condition (ii) (open-loop stability) do pose restrictions on the set of processes for which the foregoing algorithm is applicable. It is the goal of this chapter to relax restriction (i). Restriction (ii) could be relaxed if a closed-loop observer were appropriately incorporated in the control structure instead of the open-loop model simulation; this will be the subject of a future communication.

5.4 Model-algorithmic control with nonminimum-phase compensation using synthetic outputs

If a nonlinear system of the form (1) is nonminimum-phase, the model-algorithmic controller may induce closed-loop instability. In order to formulate a control

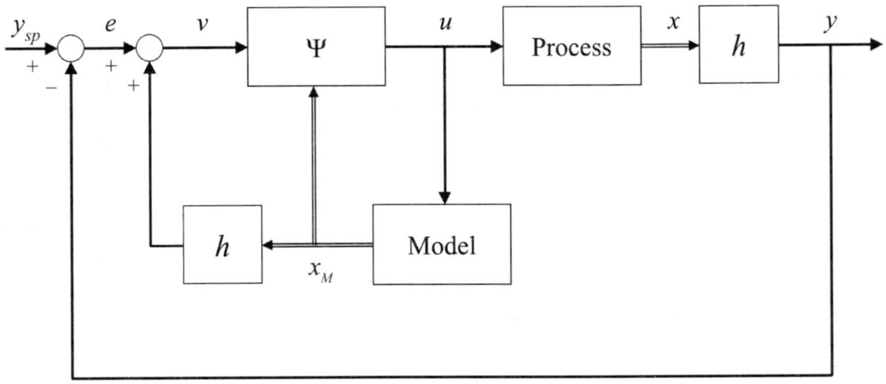

Figure 5.1 Structure of the model-algorithmic controller

algorithm that performs nonminimum-phase compensation, consider the auxiliary output maps $h_1^*(x), \ldots, h_m^*(x)$ of relative orders $r_1^*, \ldots, r_m^*$ with the following properties:

(i) The delay-free part of the system:

$$\begin{aligned} x(k+1) &= \Phi[x(k), u(k)] \\ y^*(k) &= h^*[x(k)] \end{aligned} \tag{19}$$

is locally hyperbolically minimum-phase.

(ii) The outputs $y_i = h_i(x)$ and $y_i^* = h_i^*(x)$ are statically equivalent in the sense that

$$h_i(x) = h_i^*(x) \quad \forall x \in E$$

for $i = 1, \ldots, m$.

If such output maps can be found, controlling the output y_i to a constant setpoint can be accomplished by controlling the output y_i^* to exactly the same constant setpoint. Utilising the MAC framework outlined in the previous section, future changes of the output y_i^* can be predicted according to:

$$\begin{aligned} y_{iM}^*(k+1) - y_{iM}^*(k) &= h_i^{*1}[x_M(k)] - h_i^*[x_M(k)] \\ y_{iM}^*(k+2) - y_{iM}^*(k) &= h_i^{*2}[x_M(k)] - h_i^*[x_M(k)] \\ &\vdots \\ y_{iM}^*(k+r_i^*-1) - y_{iM}^*(k) &= h_i^{*r_i^*-1}[x_M(k)] - h_i^*[x_M(k)] \\ y_{iM}^*(k+r_i^*) - y_{iM}^*(k) &= h_i^{*r_i^*-1}\{\Phi[x_M(k), u(k)]\} - h_i^*[x_M(k)] \end{aligned} \tag{20}$$

where, as before, the subscript M denotes values obtained from on-line simulation of the process model. The 'closed-loop' predictions of the auxiliary output are obtained by adding the changes to the on-line estimate $\hat{y}_i^*(k) = y_i(k) + h_i^*[x_M(k)] - h_i[x_M(k)]$:

$$\begin{aligned} \hat{y}_i^*(k+1) &= y_i(k) + h_i^{*1}[x_M(k)] - h_i[x_M(k)] \\ \hat{y}_i^*(k+2) &= y_i(k) + h_i^{*2}[x_M(k)] - h_i[x_M(k)] \\ &\vdots \\ \hat{y}_i^*(k+r_i^*-1) &= y_i(k) + h_i^{*r_i^*-1}[x_M(k)] - h_i[x_M(k)] \\ \hat{y}_i^*(k+r_i^*) &= y_i(k) + h_i^{*r_i^*-1}\{\Phi[x_M(k), u(k)]\} - h_i[x_M(k)] \end{aligned} \tag{21}$$

where $y_i(k)$ is the actual measurement of the original process output. For each of the outputs, the response is requested to follow the reference trajectory:

$$\hat{y}_i^*(k+r_i^*) = (1-\alpha_i)y_{isp} + \alpha_i \hat{y}_i^*(k+r_i^*-1) \tag{22}$$

where the subscript *sp* denotes the output setpoint and α_i is a tunable scalar parameter. Combining (21) and (22), so that the 'closed-loop' output prediction matches the reference trajectory, gives a set of nonlinear algebraic equations which depend on the manipulated input vector:

$$h_1^{*r_1^*-1}\{\Phi[x_M(k), u(k)]\} = (1-\alpha_1)\{y_{1sp} - y_1(k) + h_1[x_M(k)]\} + \alpha_1 h_1^{*r_1^*-1}[x_M(k)]$$
$$\vdots$$
$$h_m^{*r_m^*-1}\{\Phi[x_M(k), u(k)]\} = (1-\alpha_m)\{y_{msp} - y_m(k) + h_m[x_M(k)]\} + \alpha_m h_m^{*r_m^*-1}[x_M(k)] \quad (23)$$

As long as the characteristic matrix with respect to the y_i^* outputs is nonsingular, the nonlinear algebraic equations are locally solvable via the implicit function theorem. The corresponding implicit function defined as the solution to (23) is denoted by

$$u(k) = \Psi^*\{x_M(k), y_{sp} - y(k) + h[x_M(k)]\} \quad (24)$$

With the model states x_M obtained by simulating $x_M(k+1) = \Phi[x_M(k), u(k)]$ on-line, the resulting model-algorithmic controller can be expressed as:

$$\begin{aligned} x_M(k+1) &= \Phi(x_M(k), \Psi^*\{x_M(k), y_{sp} - y(k) + h[x_M(k)]\}) \\ u(k) &= \Psi^*\{x_M(k), y_{sp} - y(k) + h[x_M(k)]\} \end{aligned} \quad (25)$$

Under this controller, the closed-loop system becomes:

$$\begin{aligned} x_M(k+1) &= \Phi(x_M(k), \Psi^*\{x_M(k), y_{sp} - y(k) + h[x_M(k)]\}) \\ x(k+1) &= \Phi(x(k), \Psi^*\{x_M(k), y_{sp} - y(k) + h[x_M(k)]\}) \\ y(k) &= h[x(k)] \end{aligned} \quad (26)$$

The structure of the closed-loop system is exactly the same as the one given earlier in Figure 5.1, with Ψ^* replacing Ψ. Also, the same local stability analysis is applicable here. The closed-loop system of (26) will be locally asymptotically stable if the reference trajectories are stable $(0 \leq \alpha_i < 1)$ and the open-loop dynamics $x(k+1) = \Phi[x(k), u(k)]$ is locally hyperbolically stable.

5.5 Construction of statically equivalent outputs with pre-assigned transmission zeros

To be able to apply the control algorithm of the previous section, it is necessary to develop a procedure for calculating the synthetic outputs $y_i^* = h_i^*(x), i = 1, \ldots, m$, in order to satisfy the conditions:

(i) The process dynamics, with $y_i^* = h_i^*(x),\ i = 1, \ldots, m$, as outputs, is locally hyperbolically minimum-phase.
(ii) $y_i^* = h_i^*(x)$ is statically equivalent to $y_i = h_i(x)$.

The first step in the derivation will be the construction of $(n-m)$ independent output maps that vanish on the equilibrium manifold. Using these $(n-m)$ vanishing output maps, m outputs $h_i^*(x)$, $i=1,\ldots,m$, will be constructed so that they are statically equivalent to $h_i(x)$, $i=1,\ldots,m$, and depend on a number of arbitrary weighting parameters. Finally, the weighting parameters will be selected to place the transmission zeros of the linearised system at pre-assigned locations which will realise the local minimum-phase requirement.

5.5.1 Construction of independent functions which vanish on the equilibrium manifold

Consider a discrete-time system of the form (1). Because of assumption (A1) in Section 5.2, it is always possible to rotate the indices so that the last m rows of $(\partial\Phi/\partial u)(x,u)$ are linearly independent, i.e. the $m\times m$ matrix

$$\Gamma(x,u)=\begin{bmatrix}\frac{\partial\Phi_{n-m+1}}{\partial u_1}(x,u) & \ldots & \frac{\partial\Phi_{n-m+1}}{\partial u_m}(x,u)\\ \vdots & & \vdots\\ \frac{\partial\Phi_n}{\partial u_1}(x,u) & \ldots & \frac{\partial\Phi_n}{\partial u_m}(x,u)\end{bmatrix} \tag{27}$$

is invertible. Then, considering the equilibrium equations:

$$\begin{aligned}\Phi_1(x,u)&=x_1\\ &\vdots\\ \Phi_{n-m}(x,u)&=x_{n-m}\\ \Phi_{n-m+1}(x,u)&=x_{n-m+1}\\ &\vdots\\ \Phi_n(x,u)&=x_n\end{aligned} \tag{28}$$

it follows that the last m equations are locally solvable for u in the sense of the implicit function theorem. Therefore, there is a unique locally defined function

$$u=\Omega(x) \tag{29}$$

which solves the last m equations. Then, substituting $u=\Omega(x)$ into the first $(n-m)$ equations of (28), it follows that:

$$\begin{aligned}\Phi_1(x,\Omega(x))&=x_1\\ &\vdots\\ \Phi_{n-m}(x,\Omega(x))&=x_{n-m}\end{aligned} \tag{30}$$

at every equilibrium point. The above are $(n-m)$ eliminants of the m input variables $u_1, \ldots, u_m$ from the equilibrium equations. Thus, the functions defined by

$$\mathcal{N}_j(x) = \Phi_j(x, \Omega(x)) - x_j \quad j = 1, \ldots, n-m$$

will vanish on the equilibrium manifold by construction. Moreover, as will be seen in the proposition that follows, the above functions define a set of linearly independent scalar fields under an appropriate regularity condition.

Definition 3: *For a dynamic system of the form (1) that satisfies assumption (A1), denote by E its equilibrium set*

$$E = \{x \in X | \exists u \in U \text{ satisfying } \Phi(x, u) = x\}$$

A point $x^o \in E$ will be called a regular point of the equilibrium set if

$$\det\left[\frac{\partial \Phi}{\partial x}(x^o, u^o) - \mathrm{I}_n\right] \neq 0 \tag{31}$$

where u^o is such that $\Phi(x^o, u^o) = x^o$.

Proposition 1: *For a dynamic system of the form (1), such that the matrix $\Gamma(x, u)$ is invertible, denote by $u = \Omega(x)$ the implicit function defined as the solution of the last* m *equations of (28). Then, the functions*

$$\mathcal{N}_j(x) = \Phi_j(x, \Omega(x)) - x_j \quad j = 1, \ldots, n-m \tag{32}$$

are well defined and have the following properties:

(i) $\mathcal{N}_j(x) = 0, \quad j = 1, \ldots, n-m, \quad \forall x \in E$
(ii) $\mathcal{N}_j(x)$, $j = 1, \ldots, n-m$, *form a set of scalar fields which are linearly independent in the vicinity of every regular point of E.*

The proof of the above proposition is given in the Appendix (Section 5.9.1).

Remark 1: It should be noted that the functions $\mathcal{N}_j(x)$ of (32) are not the only functions which vanish on the equilibrium manifold. Any $(n-m)$ independent eliminant relations from (28) will generate $(n-m)$ independent functions that vanish at equilibrium.

Remark 2: Because the function $\Phi(x, u)$ is, in general, a nonlinear function of **u**, the functions $\mathcal{N}_j(x)$, $j = 1, \ldots, n-m$, must be computed numerically over a fine grid of points x which extend over a large enough region in $\Re^n$ to encompass where the state trajectories are expected to lie. This computational task could be easily bypassed if the discrete-time system (1) arises from the time-discretisation of an affine continuous-time system of the form (2). Because every approximate numerical method provides an exact description of the system's equilibrium, the equilibrium sets of the discrete-time and continuous-time system coincide. Therefore, the functions $\mathcal{N}_j(x)$ can be computed from the continuous-time description, and due to the affine dependence on u, the result can be expressed in closed form. This calculation was actually performed in earlier work of the authors [13] on nonminimum-phase compensation for continuous-time systems of the form (2). Using similar assumptions and methodology to the one developed here for discrete-time systems, the following result was derived:

$$\mathcal{N}_j(x) = f_j(x) - [g_{1,j}(x) \ldots g_{m,j}(x)]\mathcal{G}^{-1}(x)\begin{bmatrix} f_{n-m+1}(x) \\ \vdots \\ f_n(x) \end{bmatrix} \quad j = 1, \ldots, n-m \tag{33}$$

where

$$\mathcal{G}(x) = \begin{bmatrix} g_{1,n-m+1}(x) & \ldots & g_{m,n-m+1}(x) \\ \vdots & & \vdots \\ g_{1,n}(x) & \ldots & g_{m,n}(x) \end{bmatrix}$$

5.5.2 *A class of statically equivalent outputs*

Once a set of independent functions $\mathcal{N}_j(x)$, $j = 1, \ldots, n-m$, that vanish on the equilibrium set has been constructed, one can consider the class of output maps:

$$\begin{aligned} h_1^*(x) &= h_1(x) + \sum_{j=1}^{n-m} \lambda_{1,j}(x)\mathcal{N}_j(x) \\ &\vdots \\ h_m^*(x) &= h_m(x) + \sum_{j=1}^{n-m} \lambda_{m,j}(x)\mathcal{N}_j(x) \end{aligned} \tag{34}$$

where $\lambda_{i,j}(x)$ are arbitrary weighting functions. The above output maps $h_i^*(x)$, $i = 1, \ldots, m$, will be statically equivalent to $h_i(x)$, $i = 1, \ldots, m$, in the sense that

$$h_i^*(x) = h_i(x) \quad \forall x \in E$$

for any weighting functions $\lambda_{i,j}(x)$.

5.5.3 Assignment of transmission zeros

This subsection will outline a procedure for selection of m output functions out of the class (34) such that the linear approximation of:

$$x(k+1) = \Phi[x(k), u(k)]$$
$$y^*(k) = h^*[x(k)]$$

has prescribed transmission zeros. To this end, it is sufficient to consider a subclass of (34) with constant weights $\lambda_{i,j}$:

$$\begin{aligned} h_1^*(x) &= h_1(x) + \sum_{j=1}^{n-m} \lambda_{1,j} \mathcal{N}_j(x) \\ &\vdots \\ h_m^*(x) &= h_m(x) + \sum_{j=1}^{n-m} \lambda_{m,j} \mathcal{N}_j(x) \end{aligned} \tag{35}$$

or, in a more compact notation,

$$h^*(x) = h(x) + \Lambda \mathcal{N}(x)$$

where

$$\Lambda = \begin{bmatrix} \lambda_{11} & \cdots & \lambda_{1,n-m} \\ \vdots & & \vdots \\ \lambda_{m1} & \cdots & \lambda_{m,n-m} \end{bmatrix} \text{ and } \mathcal{N}(x) = \begin{bmatrix} \mathcal{N}_1(x) \\ \vdots \\ \mathcal{N}_{n-m}(x) \end{bmatrix}$$

To find the necessary values of Λ that will place the transmission zeros at desired locations, it is first necessary to introduce the following lemma.

Lemma 1: *Let A, B, C, P, Q be $n \times n$, $n \times m$, $m \times n$, $l \times l$, $m \times l$ matrices, respectively, such that P and A do not have common eigenvalues and (Q, P) is an observable pair. Also, let S be the $n \times l$ matrix which satisfies*

$$SP - AS = BQ \tag{36}$$

If $CS = 0$, then every eigenvalue of P *is a transmission zero of the dynamic system:*

$$\begin{aligned} x(k+1) &= Ax(k) + Bu(k) \\ y(k) &= Cx(k) \end{aligned} \tag{37}$$

The proof is given in the Appendix (Section 5.9.2).

The above lemma is now applied to the linearisation of (19) with output $y^*(k) = h^*[x(k)] = h[x(k)] + \Lambda\mathcal{N}[x(k)]$ around the equilibrium (x_s, u_s):

$$\begin{aligned} x(k+1) &= Ax(k) + Bu(k) \\ y^*(k) &= (C + \Lambda\mathcal{N})x(k) \end{aligned} \tag{38}$$

where:

$$A = \frac{\partial\Phi[x_s, u_s]}{\partial x}, \quad B = \frac{\partial\Phi[x_s, u_s]}{\partial u}, \quad C = \frac{\partial h[x_s]}{\partial x}, \quad N = \frac{\partial\mathcal{N}[x_s]}{\partial x}$$

Applying the condition $(C + \Lambda\mathcal{N})S = 0$ to the system of (38), the $m \times (n-m)$ matrix Λ is found to be

$$\Lambda = -(CS)(NS)^{-1} \tag{39}$$

where S is the $n \times (n-m)$ matrix which solves $SP - AS = BQ$, with P, Q being $(n-m) \times (n-m)$ and $m \times (n-m)$, respectively. This leads to the following proposition.

Proposition 2: *Let P, Q be $(n-m) \times (n-m)$ and $m \times (n-m)$ matrices, respectively, such that:*

(i) *P has distinct eigenvalues.*
(ii) *P does not have any common eigenvalues with $\partial\Phi[x_s, u_s]/\partial x$.*
(iii) *(Q, P) is an observable pair.*

Also, let S be the $n \times (n-m)$ matrix that solves the equation

$$SP - \left[\frac{\partial\Phi[x_s, u_s]}{\partial x}\right]S = \left[\frac{\partial\Phi[x_s, u_s]}{\partial u}\right]Q$$

and assume that S *has full rank. Then the system:*

$$\begin{aligned} x(k+1) &= \Phi[x(k), u(k)] \\ y^*(k) &= h[x(k)] + \Lambda\mathcal{N}[x(k)] \end{aligned} \tag{40}$$

with

$$\Lambda = -\left(\left[\frac{\partial h[x_s]}{\partial x}\right]S\right)\left(\left[\frac{\partial\mathcal{N}[x_s]}{\partial x}\right]S\right)^{-1}$$

has the property that the transmission zeros of its linear approximation are exactly the eigenvalues of the matrix P.

It is important to note that choosing the location of the transmission zeros does not uniquely determine the set of values for Λ. There are $m(n-m)$ degrees of freedom within the Q matrix, leaving $(m-1)(n-m)$ extra degrees of freedom after placing the zeros. These extra degrees of freedom can be specified by optimisation. For example, one could define an objective function for the system such as

$$\mathrm{ISE}^* = \sum_{i=1}^{m} \sum_{k=0}^{\infty} \rho_i [y_{isp} - y_i(k)]^2 \Delta t \tag{41}$$

where ρ_i are pre-assigned weights on the outputs and $y_i(k)$ is the output response under the controller (25) constructed on the basis of the synthetic output of (40). For each set of adjustable controller parameters α, minimising this function over the range of the parameters of Q will yield the best values for the construction of the matrix Λ.

Remark 3: So far, nothing has been said about the choice of the set of transmission zeros for the statically equivalent synthetic output. If the discrete-time system (1) were linear, then the functions $\mathcal{N}_j(x)$ would be linear and therefore the synthetic outputs (35) would be linear. In this case, the zeros for the synthetic output become poles of the closed-loop system arising from the proposed controller. Therefore, in terms of the ISE* criterion, the optimal location of zeros would be at the zeros of (1) which are inside the unit circle, at the inverses of the zeros of (1) which are outside the unit circle, and the remaining zeros near the origin [15]. For a nonlinear discrete-time system of the form (1), this choice of zeros will only be approximately ISE*-optimal for small step changes in the vicinity of the reference equilibrium point.

5.6 Application: control of a nonminimum-phase chemical reactor

To illustrate the application of the proposed control strategy, consider a nonisothermal continuous stirred-tank reactor where a series/parallel van der Vusse reaction is taking place [16–18]:

$$\underset{(A)}{C_5H_6} \xrightarrow{k_1/(+H_2O)} \underset{(B)}{C_5H_7OH} \xrightarrow{k_2/(+H_2O)} \underset{(C)}{C_5H_8(OH)_2}$$

$$\underset{(A)}{2C_5H_6} \xrightarrow{k_3} \underset{(D)}{C_{10}H_{12}}$$

The desired product, Cyclopentenol (*B*), is being produced through the acid-catalysed electrophilic addition of water to Cyclopentadiene (*A*). Cyclopentanediol (*C*) and Dicyclopentadiene (*D*) are also being produced as unwanted side products. The reactor model consists of mole balances for species *A* and *B* and an energy balance for the reactor. Discretising these balances using Euler's method with time step Δt gives:

$$C_A(k+1) = C_A(k) + \left\{\frac{F}{V}(C_{AO} - C_A(k)) - k_1[T(k)]C_A(k) - k_3[T(k)]C_A^2(k)\right\}\Delta t$$

$$C_B(k+1) = C_B(k) + \left\{-\frac{F}{V}C_B(k) + k_1[T(k)]C_A(k) - k_2[T(k)]C_B(k)\right\}\Delta t \quad (42)$$

$$T(k+1) = T(k) - \frac{1}{\rho C_P}\{k_1[T(k)]C_A(k)\Delta H_{R1} + k_2[T(k)]C_B(k)\Delta H_{R2} + k_3[T(k)]C_A^2(k)\Delta H_{R3}\}\Delta t + \left(\frac{F}{V}(T_O - T(k)) + \frac{Q}{\rho C_P}\right)\Delta t$$

where C_A, C_B are the molar concentrations of A, B, respectively, T is the reactor temperature, F/V is the dilution rate, and Q is the rate of heat added or removed per unit volume. C_p and ρ are the heat capacity and density of the reacting mixture, respectively, and ΔH_i are the heats of reaction. The rate coefficients are dependent on the reactor temperature via the Arrhenius equation,

$$k_i(T) = k_{iO}\exp\left(\frac{E_i}{RT}\right) \quad i = 1, 2, 3 \quad (43)$$

The parameters of the system are as given in Table 5.1.

The control objective is to maintain the outputs $y_1 = T$ and $y_2 = C_B$ at their setpoints by manipulating the dilution rate, $u_1 = F/V$, and the rate of heat addition or removal per unit volume, $u_2 = Q$. Initially, the reactor is operating at a steady state of $C_{As} = 1.25\,\text{mol/l}$, $C_{Bs} = 0.90\,\text{mol/l}$ and $T_s = 407.15\,\text{K}$, which corresponds to $u_{1s} = 19.52/\text{h}$ and $u_{2s} = -451.51\,\text{kJ/(lh)}$. Around this steady state, the process is locally asymptotically stable with eigenvalues of 0.0353 and $0.6685 \pm 0.0982i$. The transmission zero of the linear approximation of the system is found to be 2.227. This indicates that the process is locally nonminimum-phase around the given steady-state due to the transmission zero lying outside the unit circle.

Since $n - m = 1$, there is only one independent function that vanishes on the equilibrium manifold. One such function can be easily generated by eliminating $u_1 = F/V$ from the first and second state equations at equilibrium:

$$\mathcal{N}(C_A, C_B, T) = -[k_1(T)C_A + k_3(T)C_A^2]C_B + (C_{AO} - C_A)[k_1(T)C_A - k_2(T)C_B] \quad (44)$$

Table 5.1 Process parameters

$C_{A0} = 5.0\,\text{mol/l}$	$k_{10} = 1.287 \times 10^{12}\,/\text{h}$
$T_0 = 403.15\,\text{K}$	$k_{20} = 1.287 \times 10^{12}\,/\text{h}$
$\rho = 0.9342\,\text{kg/l}$	$k_{30} = 9.403 \times 10^9\,\text{l/(mol h)}$
$C_p = 3.01\,\text{kJ/(kg K)}$	$E_1/R = -9758.3\,\text{K}$
$\Delta H_1 = 4.20\,\text{kJ/mol}$	$E_2/R = -9758.3\,\text{K}$

Accordingly, statically equivalent outputs for the minimum-phase system can be constructed as follows:

$$\begin{aligned} y_1^*(k) &= T(k) + \lambda_1 \mathcal{N}[C_A(k), C_B(k), T(k)] \\ y_2^*(k) &= C_B(k) + \lambda_2 \mathcal{N}[C_A(k), C_B(k), T(k)] \end{aligned} \tag{45}$$

Placing the transmission zero at 0.449, which is the inverse of 2.227, the solution to (39) is calculated to be:

$$\begin{aligned} \lambda_1 &= -\left(\frac{1.6369 + 0.006819\ q_2/q_1}{65.7820 + 0.5992\ q_2/q_1}\right) \\ \lambda_2 &= \left(\frac{0.2940 + 0.002679\ q_2/q_1}{65.7820 + 0.5992\ q_2/q_1}\right) \end{aligned} \tag{46}$$

It is apparent that the extra degree of freedom lies in the ratio q_2/q_1 of the entries of the matrix $Q = \begin{bmatrix} q_1 \\ q_2 \end{bmatrix}$. The weights λ_1 and λ_2 are plotted against the ratio $-q_2/q_1$ in Figure 5.2. To find the optimum values of q_2/q_1, the following objective function is defined:

$$\text{ISE}^* = \sum_{k=0}^{\infty} \left(\rho_1 [T_{sp} - T(k)]^2 + \rho_2 [C_{Bsp} - C_B(k)]^2\right) \Delta t \tag{47}$$

and is minimised subject to a step change in the setpoint of $y_2 = C_B$ for a given set of controller parameters. For this case study, the values of $\alpha_1 = \alpha_2 = 0.7$ were chosen as representative values for the simulations. The ISE* profiles for these controller parameters with varying output weights and ratio values are given in Figure 5.3.

It is clearly seen that there exists a $-q_2/q_1$ value that minimises the ISE* for each output weighting ρ_1/ρ_2 at $\alpha_1 = \alpha_2 = 0.7$. Choosing equal weights in the objective function $\rho_1/\rho_2 = 1$, the optimal ratio is – 240.0801. The corresponding λ values are $\lambda_1 = -2.5508 \times 10^{-6}$ and $\lambda_2 = 0.004472$. In a more general manner, the optimal values can be found for the range of controller parameters. Given a desired output weighting value, the ratio $-q_2/q_1$ can be directly found according to Figure 5.4 for each set of controller parameters.

A step change in the setpoint of y_2 was simulated to test the performance of the proposed controller. The setpoint of y_2 was changed from 0.9 mol/l to 1.0 mol/l, while maintaining the setpoint of y_1 at 407.15 K. Figure 5.5 shows the closed-loop profiles for both the original output $y_2 = C_B$ and the synthetic output y_2^*. It is observed that C_B shows an initial inverse response due to the unstable zero dynamics of the system but reaches the new steady state in under 0.2 h. This shows that the model-algorithmic controller constructed from a system with synthetic outputs is stable and yields zero steady-state error. As the construction of the control algorithm dictates, the synthetic output follows the pre-assigned trajectory of (22). The outputs do not match during the transient period but they converge at

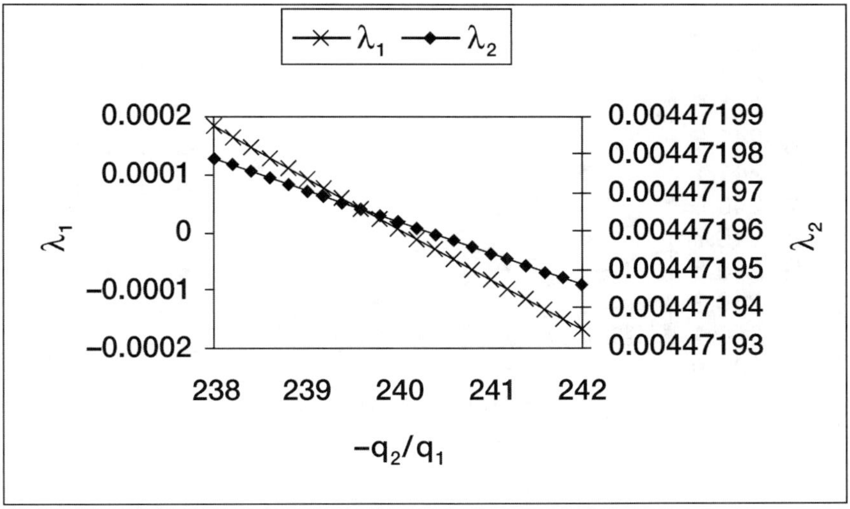

Figure 5.2 Synthetic output weights as functions of the ratio $-q_2/q_1$

the new operating steady state. Figure 5.6 shows the closed-loop profiles of $y_1 = T$ and the synthetic output y_1^*. It is seen that the controller is able to control the reactor temperature at the requested setpoint of 407.15 K without any deviations. The controller essentially maintains decoupling despite a controller design based on synthetic outputs. Since the weight in y_1^* is $\lambda_1 = -2.5508 \times 10^{-6}$, the $\mathcal{N}(x)$ function will add negligible value to the output T. The open-loop simulation of the

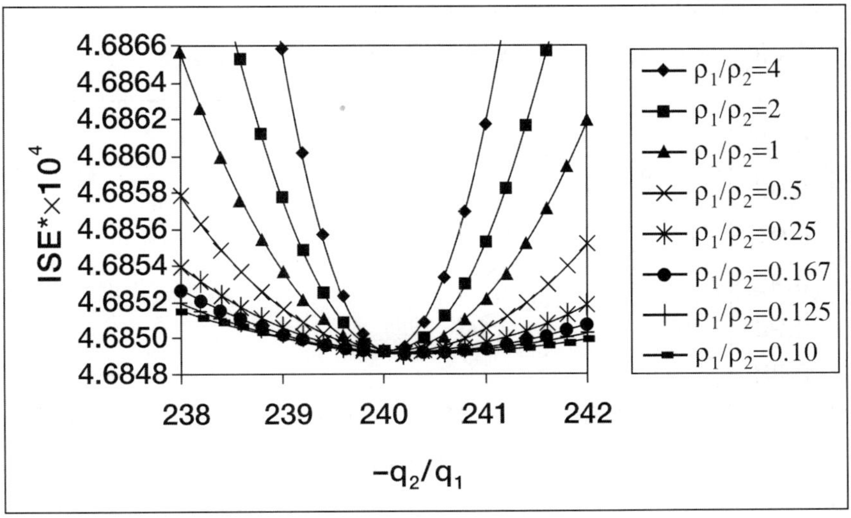

Figure 5.3 ISE profiles for $\alpha = 0.7$*

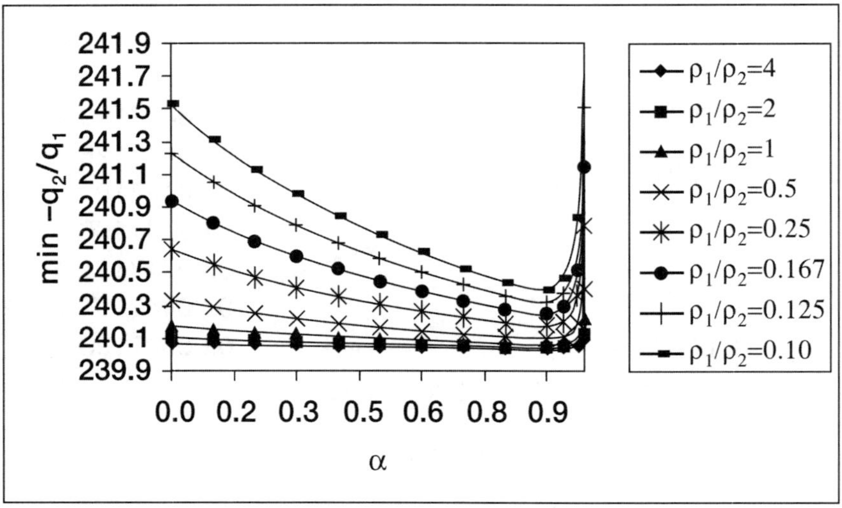

Figure 5.4 Optimal $-q_2/q_1$ values for a range of controller parameters

system of (42) shows that only the output C_B exhibits the initial inverse response. Therefore, the method will produce larger weights for the outputs with inverse responses since the remaining outputs can closely follow the requested trajectories of the model-algorithmic controller. Figure 5.7 shows the manipulated input profiles associated with the output profiles. The dilution rate increases to supply the

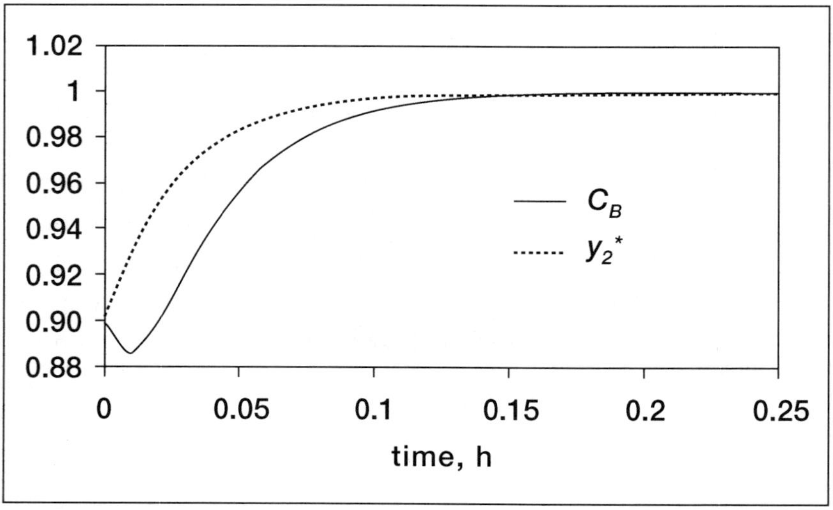

Figure 5.5 Closed-loop profiles of C_B and y_2^*

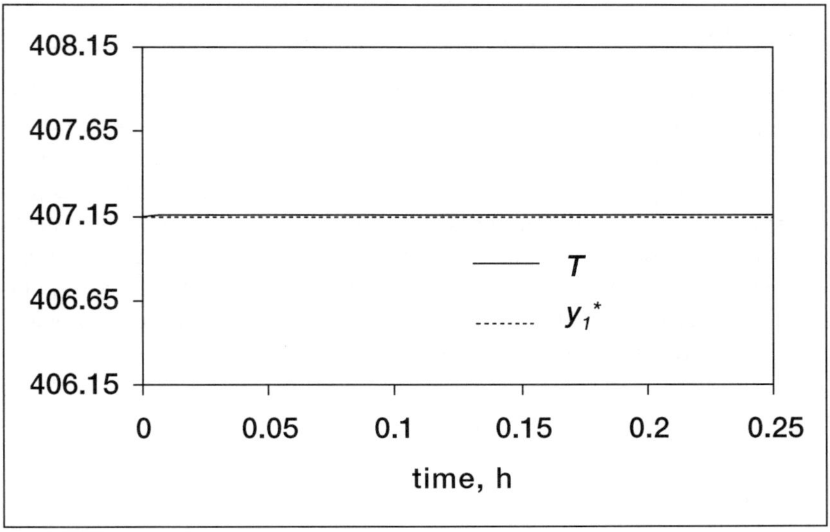

Figure 5.6 Closed-loop profiles of T and y_1^*

additional reactant needed to raise the conversion of *A* to *B*. This increase in the reaction rate will produce more heat, which requires the additional cooling seen in the profile. Overall, the proposed controller utilising synthetic outputs leads to excellent setpoint tracking and regulatory behaviour of the closed-loop system.

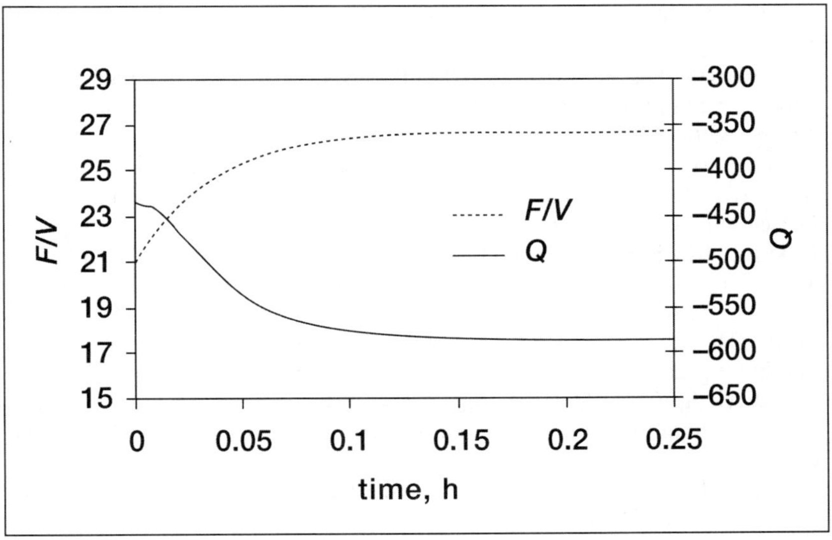

Figure 5.7 Manipulated input profiles

5.7 Conclusion

A model-algorithmic controller that utilises synthetic outputs which make the system minimum-phase is presented. A procedure is proposed for the construction of the synthetic outputs, which guarantees static equivalence to the actual process outputs and has pre-assigned transmission zeros. The proposed concepts are illustrated with a nonminimum-phase chemical reactor where a series/parallel reaction is taking place. It is shown through simulation results that the proposed controller leads to excellent setpoint tracking and regulatory behaviour in closed-loop, and therefore provides an effective solution to the difficult problem of controlling discrete-time nonminimum-phase systems.

5.8 References

1 RICHALET, J., RAULT, A., TESTUD, J.L., and PAPON, J.: 'Model predictive heuristic control: application to industrial processes', *Automatica*, 1978, **14**, pp. 413–28

2 MEHRA, R.K., and ROUHANI, R.: 'Theoretical considerations on model algorithmic control for nonminimum phase systems'. Proceedings of ACC, 1980, TA8-B

3 MEHRA, R.K., ROUHANI, R., and PRALY, R.: 'New theoretical developments in multivariable predictive algorithmic control'. Proceedings of ACC, 1980, FA9-B

4 SOROUSH, M., and KRAVARIS, C.: 'Discrete-time nonlinear controller synthesis by input/output linearization', *AIChE J.*, 1992, **38**, pp. 1923–45

5 SOROUSH, M., and KRAVARIS, C.: 'MPC formulation of GLC', *AIChE J.*, 1996, **42**, pp. 2377–81

6 KRAVARIS, C., and DAOUTIDIS, P.: 'Nonlinear state feedback control of 2nd-order nonminimum-phase nonlinear systems', *Comput. Chem. Eng.*, 1990, **14**, pp. 439–49

7 DOYLE III, F.J., ALLGÖWER, F., OLIVEIRA, S., GILLES, E., and MORARI, M.: 'On nonlinear systems with poorly behaved zero dynamics'. Proceedings of ACC, 1992, pp. 2571–5

8 DOYLE III, F.J., ALLGÖWER, F., and MORARI, M.: 'A normal form approach to approximate input-output linearisation for maximum phase nonlinear SISO systems', *IEEE Trans. Autom. Control*, 1996, **41**, pp. 305–9

9 WRIGHT, R.A., and KRAVARIS, C.: 'Nonminimum-phase compensation for nonlinear processes', *AIChE J.*, 1992, **38**, pp. 26–40

10 KRAVARIS, C., DAOUTIDIS, P., and WRIGHT, R.A.: 'Output feedback control of nonminimum-phase nonlinear processes', *Chem. Eng. Sci.*, 1994, **49**, pp. 2107–22

11 KRAVARIS, C., NIEMIEC, M., BERBER, R., and BROSILOW, C.: 'Nonlinear model-based control of nonminimum-phase processes', in BERBER, R., and KRAVARIS, C. (Eds): 'Nonlinear model based process control' (Kluwer Academic Publishers, 1998), pp. 115–41

12 NIEMIEC, M., and KRAVARIS, C.: 'Nonlinear model-algorithmic control: a review and new developments', in BERBER R., and KRAVARIS C. (Eds), (Kluwer Academic Publishers, 1998), pp. 143–71
13 NIEMIEC, M., and KRAVARIS, C.: 'Controller synthesis for multivariable nonlinear nonminimum-phase processes'. Proceedings of ACC, 1998, pp. 2076–80
14 NIJMEIJER, H., and VAN DER SCHAFT, A.J.: 'Nonlinear dynamical control systems' (Springer-Verlag, Berlin, 1990)
15 ASTRÖM, K.J., and WITTENMARK, B.: 'Computer controlled systems: theory and design' (Prentice Hall, Englewood Cliffs, NJ, 1984)
VAN DE VUSSE, J.G.: 'Plug-flow-type reactor versus tank reactor', *Chem. Eng. Sci.*, 1964, **19**, pp. 994–8
16 KANTOR, J.C.: 'Stability of state feedback transformations for nonlinear systems – some practical considerations'. Proceedings of ACC, 1986, pp. 1014–16
17 ENGELL, S., and KLATT, K.U.: 'Nonlinear control of a nonminimum-phase CSTR'. Proceedings of ACC, 1993, pp. 2341–45

5.9 Appendix

5.9.1 Proof of Proposition 1

The first property is an immediate consequence of (30). To prove the second property, observe that

$$\begin{bmatrix} \frac{\partial \mathcal{N}_1}{\partial x}(x) \\ \vdots \\ \frac{\partial \mathcal{N}_{n-m}}{\partial x}(x) \end{bmatrix} = \begin{bmatrix} \frac{\partial \Phi_1}{\partial x}(x,\Omega(x)) \\ \vdots \\ \frac{\partial \Phi_{n-m}}{\partial x}(x,\Omega(x)) \end{bmatrix} + \begin{bmatrix} \frac{\partial \Phi_1}{\partial u}(x,\Omega(x)) \\ \vdots \\ \frac{\partial \Phi_{n-m}}{\partial u}(x,\Omega(x)) \end{bmatrix} \frac{\partial \Omega}{\partial x}(x) - [\mathbf{I}_{n-m}\ |0]$$

From the definition of $\Omega(x)$ and the implicit function theorem,

$$\frac{\partial \Omega}{\partial x}(x) = -\begin{bmatrix} \frac{\partial \Phi_{n-m+1}}{\partial u}(x,\Omega(x)) \\ \vdots \\ \frac{\partial \Phi_{n}}{\partial u}(x,\Omega(x)) \end{bmatrix}^{-1} \left(\begin{bmatrix} \frac{\partial \Phi_{n-m+1}}{\partial x}(x,\Omega(x)) \\ \vdots \\ \frac{\partial \Phi_{n}}{\partial x}(x,\Omega(x)) \end{bmatrix} - [0 \mid \mathbf{I}_m] \right)$$

Combining the above expressions, it follows that

$$\begin{bmatrix} \frac{\partial \mathcal{N}_1}{\partial x}(x) \\ \vdots \\ \frac{\partial \mathcal{N}_{n-m}}{\partial x}(x) \end{bmatrix} = \left[\mathrm{I}_{n-m} \;\middle|\; - \begin{bmatrix} \frac{\partial \Phi_1}{\partial u}(x,\Omega(x)) \\ \vdots \\ \frac{\partial \Phi_{n-m}}{\partial u}(x,\Omega(x)) \end{bmatrix} \begin{bmatrix} \frac{\partial \Phi_{n-m+1}}{\partial u}(x,\Omega(x)) \\ \vdots \\ \frac{\partial \Phi_n}{\partial u}(x,\Omega(x)) \end{bmatrix}^{-1} \right] \begin{bmatrix} \begin{bmatrix} \frac{\partial \Phi_1}{\partial x}(x,\Omega(x)) \\ \vdots \\ \frac{\partial \Phi_{n-m}}{\partial x}(x,\Omega(x)) \end{bmatrix} - [\mathrm{I}_{n-m} \mid 0] \\ \begin{bmatrix} \frac{\partial \Phi_{n-m+1}}{\partial x}(x,\Omega(x)) \\ \vdots \\ \frac{\partial \Phi_n}{\partial x}(x,\Omega(x)) \end{bmatrix} - [0 \mid \mathrm{I}_m] \end{bmatrix}$$

$$= \left[\mathrm{I}_{n-m} \;\middle|\; - \begin{bmatrix} \frac{\partial \Phi_1}{\partial u}(x,\Omega(x)) \\ \vdots \\ \frac{\partial \Phi_{n-m}}{\partial u}(x,\Omega(x)) \end{bmatrix} \begin{bmatrix} \frac{\partial \Phi_{n-m+1}}{\partial u}(x,\Omega(x)) \\ \vdots \\ \frac{\partial \Phi_n}{\partial u}(x,\Omega(x)) \end{bmatrix}^{-1} \right] \left(\frac{\partial \Phi}{\partial x}(x,\Omega(x)) - \mathrm{I}_n \right)$$

The first factor of the right-hand side has rank $n - m$. The second factor of the right-hand side is a nonsingular square matrix by assumption of regularity. Therefore, the left-hand side has rank $n - m$.

5.9.2 Proof of Lemma 1

Let p_i be an eigenvalue of P and v_i a corresponding eigenvector. Then $Pv_i = p_i v_i$ and the matrix equation $SP - AS = BQ$ implies that:

$$\begin{aligned} SPv_i - ASv_i &= BQv_i \\ \text{or } Sp_i v_i - ASv_i &= BQv_i \\ \text{or } (p_i\mathrm{I} - A)Sv_i &= BQv_i \\ \text{or } Sv_i &= (p_i\mathrm{I} - A)^{-1}BQv_i \end{aligned}$$

Multiplying by C from the left-hand side and using the condition $CS = 0$, it follows that

$$C(p_i\mathrm{I} - A)^{-1}BQv_i = 0$$

But, because (Q, P) is an observable pair, $Qv_i \neq 0$ for every eigenvector v_i of P. Therefore

$$\det[C(p_i\mathrm{I} - A)^{-1}B] = 0$$

This proves that p_i is a transmission zero of the system:

$$\begin{aligned} x(k+1) &= Ax(k) + Bu(k) \\ y(k) &= Cx(k) \end{aligned}$$

Chapter 6

Open-loop and closed-loop optimality in interpolation MPC

M. Cannon and B. Kouvaritakis

Abstract

The chapter describes efficient MPC algorithms based on univariate interpolation for stable output regulation of nonlinear systems. Two techniques for ensuring closed-loop stability are discussed: inclusion of a suitable terminal penalty term in the cost; and explicit enforcement of a convergence condition via an additional constraint in the online optimisation. An open-loop analysis shows the relationship between these approaches, and inverse optimality results are derived in order to compare their closed-loop properties. These analyses, together with numerical simulations, indicate the advantages of the convergence constraint approach over the more conventional terminal penalty approach.

6.1 Introduction

Conventional MPC algorithms employing future control variables as the degrees of freedom in the online optimisation are subject to a trade-off of performance and size of stabilisable set against computational load: short horizons are computationally less expensive but result in poor performance and small stabilisable initial condition sets. Several methods of overcoming this fundamental compromise have recently been proposed [1–4]. In References 1 and 2 only the first element of the predicted future control trajectory is optimised, and the optimal values of future controls are approximated via a gain scheduled control law designed offline. On the other hand the approach of References 3 and 4 interpolates online between control laws designed offline, and is thus able to optimise performance and handle constraints systematically over horizons greater than a single sample interval.

The current work extends the approach of References 3 and 5 to the case of nonlinear, control affine plant dynamics. A univariate online optimisation is derived by interpolating between a control law which is optimal (in a suitable sense) in the absence of constraints (but may violate system constraints and may not be stabilising) and a control law with a large associated stabilisable set (but which is suboptimal). By including these control laws in the predictions over which performance is optimised, the interpolation MPC law inherits both their desirable optimality and feasibility properties. Whereas Reference 3 considers linear systems and is therefore able to ensure stability by using an infinite prediction horizon, other techniques are needed to guarantee stability in the current context of nonlinear systems. The choice of stabilising technique has a significant effect on the performance interpolation MPC due to the use of an optimisation based on only a single degree of freedom. This chapter considers which stabilising technique is most suitable for interpolation MPC algorithms.

We compare two methods of ensuring closed-loop stability: (a) inclusion of a suitable terminal state penalty in the receding horizon cost; (b) inclusion of an artificial convergence constraint in the online optimisation. Method (a) was proposed in References 6–8 and has found most widespread acceptance in the nonlinear MPC literature [9]. Method (b) was used in References 10 and 11 in the context of linear discrete-time systems, and also in References 12 and 13 to derive MPC laws based on inverse-optimal controllers for continuous-time nonlinear systems. The approach of (a) provides a guarantee of stability if the terminal weight is chosen to be a control Lyapunov function for the constrained plant; on the other hand the artificial constraint in (b) forces a pre-specified function to act as a Lyapunov function for the closed-loop plant.

For the interpolation MPC laws based on methods (a) and (b) considered below, optimality with respect to an infinite horizon cost follows if the terminal weight in (a) or the value function in (b) corresponds to the optimal value of this cost. In practice, however, the optimal value function associated with a given infinite-horizon performance index is unknown, and by considering open-loop and inverse optimality properties we conclude that in this case method (a) is likely to result in worse performance since in general the terminal weight dominates the receding horizon cost. This conclusion is supported by the results of numerical simulations performed on an input constrained bilinear plant.

6.2 Problem statement

Consider the discrete-time system with model

$$x(k+1) = f(x(k)) + g(x(k))u(k) \tag{1}$$

where $x \in \Re^{n_x}$, $u \in \Re^{n_u}$ are the state and input variables, and f, g are Lipschitz continuous functions satisfying $f(0) = 0$, $g(0) = 0$. At each sample instant the input is subject to time-invariant constraints of the form

$$|u(k)| \leq \bar{u} \tag{2}$$

($|u|$ denotes the vector of absolute values of elements of u). Note that the approach described below can easily be extended to handle more general input and state constraints of the form $u \in U$, $x \in X$ for arbitrary closed, compact sets U, X with $0 \in U, 0 \in X$.

This chapter considers the design of efficient MPC laws with the objective of minimising a performance criterion defined by

$$J_\infty = \sum_{k=1}^{\infty} \|x(k)\|^2_{C^TQC} \tag{3}$$

Here $\|y\|_Q$ denotes the weighted norm $(y^TQy)^{1/2}$ for $Q > 0$, and $y(k) = Cx(k)$ is a vector of costed outputs of dimension n_y. The linear output map $Cx(k)$ is assumed purely for convenience; the framework for ensuring closed-loop stability described below is unaffected if this is replaced by a more general nonlinear function $y(k) = h(x(k))$.

The stage cost $\|x(k)\|^2_{C^TQC}$ is a positive semidefinite (not necessarily positive definite) function of the state $x(k)$, and consequently the closed-loop system formed by the plant (1) under a control law optimal with respect to J_∞ may not have a stable equilibrium at $x = 0$. This chapter therefore determines control laws which minimise approximately the cost of (3) subject to system constraints (2) and to the constraint that $x(k)$ remains bounded for all $k > 0$.

Note that it is usual to include terms penalising input activity of the form $\|u(k)\|^2_R$, $R > 0$, in MPC performance criteria. Historically this form of performance criterion was used in the design of LQ control laws in order to reduce control activity and thus avoid violating constraints. Here, however, input constraints are handled more systematically by incorporating (2) in an online optimisation, and consequently an input penalty term is no longer needed in the cost.

6.3 Predicted input/state trajectories

Interpolation MPC algorithms are based on an extremely efficient univariate optimisation. However, despite their low computational burden, they provide large stabilisable initial condition sets and are capable of a high degree of optimality. This is achieved by specifying the input trajectory predicted at sample instant k as a univariate interpolation between two control laws:

$$\begin{aligned} u(k+i|k) = (1-\alpha(k))u^o(x(k+i|k)) + \alpha(k)u^f(x(k+i|k)),\\ i = 0, 1, \ldots, N \end{aligned} \tag{4}$$

where the interpolation variable $\alpha(k) \in [0, 1]$ is taken to be constant over the prediction horizon N, and the state predictions $x(k+i|k)$ are governed by the plant dynamics:

$$x(k+i+1|k) = f(x(k+i|k)) + g(x(k+i|k))u(k+i|k) \tag{5}$$

The feedback laws u^o and u^f are designed offline: u^o is chosen to be optimal with respect to J_∞ in the absence of constraints; u^f is a feedback law which is stabilising for the constrained plant over as large a set of initial conditions as possible. Whenever feasible, the choice $\alpha(k) = 0$ therefore recovers the 'unconstrained optimal' control law u^o, and hence yields optimal predicted performance with respect to J_∞. On the other hand, the possibility of choosing $\alpha(k) = 1$ endows predicted trajectories with the same stability properties as the 'feasible' control law u^f.

In the following two subsections we describe methods for designing u^o and u^f. Optimality and closed-loop stability properties of MPC laws based on the predictions of (4) are then discussed in Sections 6.4 and 6.5.

6.3.1 Unconstrained optimal control law u^o

Computation of a globally valid optimal control law for the plant (1) with respect to the infinite horizon cost J_∞ can be nontrivial, even when system constraints are ignored. However we consider here the simpler problem of determining a feedback law u^o which is locally optimal for the unconstrained plant within some region of state space containing the origin. Provided the plant satisfies a controllability condition, the computation of u^o then reduces to the solution of a set of algebraic equations.

First consider the simplest case in which it is possible to construct a local control law for $u(k)$ which forces the stage cost $\|x(k+1)\|^2_{C^TQC}$ at the next sample instant to be zero. This is equivalent to the following assumption.

Assumption 1: There exists an open set S, with $0 \in S$, such that $\text{Rank}(Cg(x)) = n_y$ for all $x \in S$.

If Assumption 1 holds, then the costed output $Cx(k+1)$ is an invertible function of the input $u(k)$ whenever $x(k) \in S$, and it follows that $\|x(k+1)\|_{C^TQC} = 0$ if and only if $u(k)$ satisfies

$$Cg(x(k))u(k) = -Cf(x(k)) \tag{6}$$

The corresponding solution for u is given by $u(k) = u^o(x(k))$, where

$$u^o(x(k)) = -(Cg(x))^+ Cf(x) \tag{7}$$

Here $(Cg(x))^+$ is the inverse of the matrix $Cg(x)$ if $n_y = n_u$, and, noting that Assumption 1 requires that $n_y \leq n_u$, is otherwise defined as a right inverse of $Cg(x)$ (e.g. corresponding to the least squares solution to (6) for $u(k)$). The following lemma summarises this argument.

Lemma 1: Under Assumption 1, the cost J_∞ is minimised along trajectories of (1) under the control law $u(k) = u^o(x(k))$ given by (7), provided $x(k)$ lies in S for all $k \geq 0$.

Proof. The control law of (7) solves (6) and therefore yields the minimum cost $J_\infty = \|x(0)\|^2_{C^TQC}$.

Remark 1: *Assumption 1 is restrictive as it requires that each element of $Cx(k+1)$ depends explicitly on $u(k)$. In practice the deadtime between a change in u and the associated output response is likely to be different for the individual elements of the output vector. However, the approach described above for determining u^o has a straightforward extension to this more general case which we now briefly describe. Assume that the plant (1) with output $Cx(k)$ has relative degree $r = [r_1 \cdots r_{n_y}]^T$ (where r_i is the deadtime associated with y_i, the ith element of Cx) in an open set S containing the origin. This is equivalent to the conditions that, for all $x(k) \in S$: (a) for each i, $y_i(k+j)$ is independent of $u(k)$ for $j = 1, \ldots, r_i - 1$; and (b) $\underset{\rightarrow}{y}(k) = [y_1(k+r_i) \cdots y_{n_y}(k+r_{n_y})]^T$ satisfies*

$$\underset{\rightarrow}{y}(k) = \phi(x(k), u(k)), \mathrm{Rank}\left(\frac{\partial \phi(x,u)}{\partial u}\right) = n_y \tag{8}$$

This constitutes a relaxation of Assumption 1 and, by a similar argument to Lemma 1, it follows from (8) that the minimum value of J_∞ is obtained under the control law $u(k) = u^o(x(k))$ defined by the solution of $\phi(x(k), u^o) = 0$ if $x(k) \in S$ for all $k \geq 0$.

Remark 2: *Note that the rank condition of Assumption 1 (or more generally (8)) together with the assumption that f, g are Lipschitz continuous ensures that u^f is a Lipschitz continuous function of $x \in S$.*

The unconstrained optimal feedback law u^o derived above is a form of feedback linearising control law (see, for example, Reference 14). This is due to the dependence of J_∞ on the output sequence $Cx(k)$ alone, and as a consequence:

(a) u^ocan only be optimal in a neighbourhood of $x = 0$;
(b) the closed loop system under u^o may be unstable at $x = 0$ (e.g. if the zero dynamics of the plant are unstable at $x = 0$ [14]).

These shortcomings, together with the effects of system constraints (2) which were neglected in the design of u^o, limit the applicability of u^o in practice. In the current context of interpolation MPC, however, they are to a large extent overcome by the inclusion of the stabilising feedback law u^f in the predictions (4), which has the effect of forcing the plant state into the region of state space on which u^o has desirable optimality properties.

6.3.2 Feasible control law u^f

The feedback law u^f provides the means for ensuring the closed-loop stability of an MPC strategy based on the predictions of (4). The primary design requirement for u^f is therefore that it should stabilise the constrained plant for all initial conditions in a desired operating region. On the other hand, u^f should also give some guarantee of closed-loop performance with respect to J_∞ in order to limit the worst-case performance of the MPC algorithm. These requirements can be conflicting since good performance in the presence of constraints is often obtained only at the expense of a small stabilisable set. However, by restricting u^f to be a linear feedback law and replacing the plant dynamics by a representative set of linear models it is possible to balance these requirements systematically via an offline optimisation, as we now briefly describe.

Let $u^f(x) = Kx$ and consider the ellipsoidal set $\{x : \|x\|_P^2 \leq \gamma\}$ for $P > 0$ and $\gamma > 0$. It is easy to show that the maximum value of $|K_j x|$ (where K_j is the jth row of K) for all x in this ellipsoidal set is $\gamma K_j P^{-1} K_j^T$. Therefore $u^f(x)$ satisfies input constraints (2) whenever $\|x\|_P^2 \leq \gamma$ if and only if K, P and γ satisfy

$$\gamma K_j P^{-1} K_j^T \leq \bar{u}_j^2, \quad j = 1, \ldots, n_u \tag{9}$$

If, in addition to (9), the following condition holds for all x satisfying $\|x\|_P^2 \leq \gamma$:

$$\|f(x) + g(x)Kx\|_P^2 - \|x\|_P^2 \leq -\|f(x) + g(x)Kx\|_{C^TQC}^2 \tag{10}$$

then it is clear that, starting from any initial condition in the ellipsoidal set $\{x : \|x\|_P^2 \leq \gamma\}$, the control law $u(k) = u^f(x(k))$ satisfies (2) and the corresponding closed-loop state trajectory remains within this ellipsoid at all times $k \geq 0$. Thus (9)–(10) together ensure that $\{x : \|x\|_P^2 \leq \gamma\}$ is a stabilisable set for the closed-loop system formed by (1) under the feedback law u^f.

The following lemma shows that condition (10) also provides an upper bound on the cost J_∞ along closed-loop trajectories under u^f.

Lemma 2: If (9)–(10) hold whenever $\|x\|_P^2 \leq \gamma$, then the cost of (3) satisfies

$$J_\infty \leq \gamma \tag{11}$$

along closed-loop trajectories of (1) under u^f with initial condition $\|x(0)\|_P^2 \leq \gamma$.

Proof: Setting $x = x(k)$ in (10) and summing over $k = 0, 1, \ldots,$ we have

$$\begin{aligned} J_\infty &\leq \|x(0)\|_P^2 - \lim_{k\to\infty} \|x(k)\|_P^2 \\ &\leq \|x(0)\|_P^2 \end{aligned}$$

which implies the bound (11).

A small value for γ is therefore desirable in order to minimise the worst case performance bound given by (11). Reducing γ for given P causes a reduction in the size of the ellipsoid $\{x : \|x\|_P^2 \leq \gamma\}$, however, and to determine the parameters K, P, γ we therefore minimise a functional of the form [5]:

$$\log(\det(P/\gamma)) + c_\gamma \gamma \tag{12}$$

(where c_γ is a constant) subject to constraints which ensure satisfaction of (9)–(10). Minimisation of the first term in (12) is equivalent to maximizing the volume of the ellipsoid $\{x : \|x\|_P^2 \leq \gamma\}$, and consequently the designer is able to balance the bounds on performance and size of the stabilisable set through the choice of the coefficient c_γ.

Although the objective (12) is convex in K, P, γ, the constraints (9)–(10) will in general be nonconvex. Using the approach of Kothare *et al.* [15], sufficient conditions for satisfaction of (9)–(10) can be cast as convex linear matrix inequalities (LMIs) if the plant model (1) is represented by a set of p linear models:

$$f(x) + g(x)u \in \mathrm{Co}\{A_i x + B_i u, \quad i = 1, \ldots, p\} \tag{13}$$

where Co denotes the convex hull. The offline determination of K, P, γ then reduces to minimisation of (12) subject to LMI constraints equivalent to:

$$(A_i + B_i K)^T P (A_i + B_i K) - P \leq (A_i + B_i K)^T C^T Q C (A_i + B_i K), \; i = 1, \ldots, p \tag{14a}$$

$$\gamma K_j P K_j^T \leq \bar{u}_j^2, \quad j = 1, \ldots, n_u \tag{14b}$$

Through an appropriate parameterisation of variables (see Reference 15 for details), this problem can be solved efficiently using semidefinite programming.

Remark 3: *The assumption that f, g are Lipschitz continuous ensures that it is always possible to determine a set of linear models $\{A_i, \; B_i, \; i = 1, \ldots, p\}$ such that (13) is satisfied for x in a convex set (say T) containing the origin. The validity of replacing (9)–(10) by (14) is in this case ensured by including the additional constraint*

$$\{x : \|x\|_P^2 \leq \gamma\} \subseteq T$$

in the minimisation of (12).

6.4 Interpolation MPC algorithms

This section compares two methods for ensuring closed-loop stability of MPC algorithms based on the predictions of (4)–(5). The choice of stabilising technique has a significant effect on the performance of interpolation MPC because of the univariate nature of the online optimisation. Moreover, as mentioned above, the controller u^o yielding optimal performance with respect to J_∞ may not be stabilising at $x = 0$, and, depending on which technique is employed to guarantee closed-loop stability, this may force the interpolation algorithm to follow closely the stabilising control law u^f at the expense of performance. Below, it is shown that, in the simplest case of a prediction horizon of $N = 1$, the use of an artificial constraint which forces a prescribed function of the plant state to decrease monotonically along closed-loop trajectories is preferable to incorporating a terminal weight in the cost function minimised online.

The stability of MPC is ensured if the algorithm has the following two properties:

(a) a recursive guarantee of feasibility of the online optimisation given feasibility at start time;
(b) existence of a Lyapunov function for the closed-loop system under the optimised control law.

In order to ensure satisfaction of (a) we incorporate the constraint $\|x(k+1|k)\|_P^2 \leq \gamma$ on the terminal state prediction [16]. This forces the state at the next sample instant to lie in a set on which the feedback law u^f satisfies system constraints (2) (as well as the terminal constraint $\|x(k+1|k)\|_P^2 \leq \gamma$), and therefore ensures for each $k \geq 0$ that the predictions of (4) at time $k+1$ will include a stabilising control law.

One way to meet requirement (b) is to add a terminal weight to the finite horizon performance index in the objective to be minimised online [6–8]. In the current context of interpolation MPC with $N = 1$, this results in Algorithm T-1 below.

Algorithm T-1: *At all times $k \geq 0$, perform the following minimisation over the predictions of (4)–(5):*

$$\alpha^* = \arg\min_{\alpha(k)} \|x(k+1|k)\|_{C^TQC}^2 + \|x(k+1|k)\|_P^2 \tag{15a}$$

$$\text{subject to } |u(k|k)| \leq \bar{u} \tag{15b}$$

$$\|x(k+1|k)\|_P^2 \leq \gamma \tag{15c}$$

Implement $u(k|k)$ corresponding to $\alpha(k) = \alpha^*$ in (4).

Here the terminal weight is the term $\|x(k+1|k)\|_P^2$ in (15a). The terminal constraint (15c) ensures that the choice $\alpha(k) = 1$, corresponding to $u(k|k) = u^f(x(k))$, is feasible at times $k \geq 1$, and this provides an upper bound on the optimal value of the objective in (15a), which we denote as $J^*(k)$:

$$J^*(k) \leq \|x^f(k+1|k)\|^2_{C^TQC} + \|x^f(k+1|k)\|^2_P$$

where $x^f(k+1|k) = f(x(k)) + g(x(k))Kx(k)$. The change in the optimal value of (15a) over each sample period therefore satisfies the bound

$$\begin{aligned} J^*(k) - J^*(k-1) &\leq -\|x(k)\|^2_{C^TQC} - \|x(k)\|^2_P + \|x^f(k+1|k)\|^2_P + \|x^f(k+1|k)\|^2_{C^TQC} \\ &\leq -\|x(k)\|^2_{C^TQC} \end{aligned} \tag{16}$$

(where (10) has been used). Thus $J^*(k)$ is monotonically nonincreasing along closed-loop trajectories under Algorithm T-1, and since $x(k+1|k) = x(k+1)$ in the absence of disturbances, this implies that $x = 0$ is a stable equilibrium of the closed-loop system. Furthermore it follows from (16) that $\lim_{k\to\infty} \|x(k)\|_{C^TQC} = 0$.

The majority of recent MPC algorithms for nonlinear systems use a terminal weight as in Algorithm T-1 to ensure closed-loop stability (see, for example, the survey of Reference 9). Here we consider also an alternative approach which incorporates an additional constraint in the online optimisation instead of a terminal weight in order to ensure the desired stability properties by forcing a prespecified positive definite function of the plant state to be monotonically nonincreasing along closed-loop system trajectories. This approach appeared in References 11 and 13 for discrete- and continuous-time MPC formulations, respectively, and in the current context yields Algorithm C-1 below.

Algorithm C-1: *At all times $k \geq 0$, perform the following minimisation over the predictions of (4)–(5).*

$$\alpha^* = \arg\min_{\alpha(k)} \alpha^2(k) \tag{17a}$$

$$\text{subject to } |u(k|k)| \leq \bar{u} \tag{17b}$$

$$\|x(k+1|k)\|^2_P - \|x(k)\|^2_P \leq -\|x(k+1|k)\|^2_{C^TQC} \tag{17c}$$

$$\|x(k+1|k)\|^2_P \leq \gamma \tag{17d}$$

Implement $u(k|k)$ corresponding to $\alpha(k) = \alpha^*$ in (4).

Whenever feasible, the convergence constraint (17c) implies monotonicity of $\|x(k)\|^2_P$ and results in $\lim_{k\to\infty} \|x(k)\|_{C^TQC} = 0$. Feasibility of (17b–d) at times $k \geq 1$ is moreover ensured by the terminal constraint (17d) and the convergence property (10) of the closed-loop system under u^f. Note that minimising the objective of (17a) is equivalent to minimising the stage cost $\|x(k+1|k)\|^2_{C^TQC}$ given the control affine form of the plant dynamics.

6.4.1 Comparison of open-loop optimality

The different techniques for ensuring closed-loop stability in Algorithms T-1 and C-1 result in similar bounds on the rate at which $\|x(k)\|_P^2$ decreases along closed-loop trajectories. In fact (16), with $J^*(k-1) = \|x(k)\|^2_{C^TQC} + \|x(k)\|^2_P$ and $J^*(k) = \|x(k+1|k)\|^2_{C^TQC} + \|x(k+1|k)\|^2_P$, implies that the bound

$$\|x(k+1|k)\|_P^2 - \|x(k)\|_P^2 \leq -\|x(k+1|k)\|^2_{C^TQC} \tag{18}$$

is satisfied under the control law of Algorithm T-1. Thus both Algorithms T-1 and C-1 are effectively aiming to minimise the stage cost $\|x(k+1|k)\|^2_{C^TQC}$ while satisfying the convergence condition of (18). However, this condition is explicitly invoked as a constraint in Algorithm C-1, and it is therefore clear that Algorithm C-1 necessarily achieves a smaller stage cost than Algorithm T-1. This is shown by the following theorem, which clarifies the relationship between the two algorithms.

Theorem 1: The solution of the minimisation of Algorithm C-1 can be expressed as $\alpha^* = \alpha_{\min}(\mu^*)$, where $\alpha_{\min}(\mu)$ is defined for all $\mu \in [0,1]$ by

$$\begin{aligned} \alpha_{\min}(\mu) = \arg\min_{\alpha(k)} \|x(k+1|k)\|^2_{C^TQC} + \mu\|x(k+1|k)\|_P^2 \\ \text{subject to (17b), (17d)} \end{aligned} \tag{19}$$

and μ is defined by

$$\begin{aligned} \mu^* = \arg\min_{\mu\in[0,1]} \alpha_{\min}(\mu) \\ \text{subject to (17c), with } \alpha(\mathrm{k}) = \alpha_{\min}(\mu) \end{aligned} \tag{20}$$

where $u(k|k)$ and $x(k+1|k)$ are given by (4)–(5).

Proof: Let λ be a Lagrange multiplier for constraint (17c) in Algorithm C-1 with optimal value λ^*, and consider the function

$$L(\alpha(k),\lambda) = \|x(k+1|k)\|^2_{C^TQC} + \lambda\left\{\|x(k+1|k)\|_P^2 + \|x(k+1|k)\|^2_{C^TQC}\right\}, \quad \lambda \geq 0$$

The convexity of the minimisation of Algorithm C-1 implies that the minimisation

$$\min_{\alpha(k)} L(\alpha(k),\lambda) \quad \text{subject to (17b), (17d)}$$

for fixed λ has optimal value $\alpha(k)$ which is feasible for (17c) if $\lambda \geq \lambda^*$ and infeasible whenever $\lambda < \lambda^*$. Furthermore the objective of (19) can be written:

$$\|x(k+1|k)\|^2_{C^TQC} + \mu\|x(k+1|k)\|_P^2 = \frac{L(\alpha(k),\lambda)}{1+\lambda}; \ \mu = \frac{\lambda}{1+\lambda}$$

and the optimisation of (19) therefore yields $\alpha_{\min}(\mu)$, which is feasible for (17c) with $\alpha(k) = \alpha_{\min}(\mu)$ if and only if $\mu \geq \lambda^*/(1+\lambda^*)$. Hence the solution of (20) lies in the interval $\mu \in [0, 1]$, and $\alpha(\mu^*)$ (where $\mu^* = \lambda^*/(1+\lambda^*)$) is optimal with respect to the minimisation of Algorithm C-1 as a result of the optimality of λ^*.

The minimisation of (19) for $\mu = 1$ is identical to the optimisation problem in Algorithm T-1, and from (18) it follows that $\alpha(k) = \alpha_{\min}(1)$ satisfies the convergence constraint (17c). It is therefore clear from (20) that Algorithm C-1 yields a smaller optimal value for the interpolation variable α than Algorithm T-1, and hence also a smaller stage cost $\|x(k+1|k)\|^2_{C^TQC}$ (see Figure 6.1).

Theorem 1 provides an insight into the relationship between the terminal weight and convergence constraint methods of ensuring closed-loop stability. From (19) it can be seen that the convergence constraint (17c) in Algorithm C-1 has the effect of adding to the objective (17a) a terminal weight of the same form as that employed in the minimisation of Algorithm T-1, but scaled by a factor $\mu \leq 1$. As a result of the convexity of (19), the minimisation of (20) is equivalent to reducing μ to the minimum value for which the convergence constraint (17c) is satisfied. This scaling process reduces the influence of the terminal weight on the objective function, and is particularly advantageous in MPC algorithms employing short prediction horizons since the terminal weight can otherwise dominate the objective in the online optimisation problem.

6.4.2 Closed-loop optimality properties

It is clearly possible to find examples for which an initial control $u(0)$ yielding a small value of $\|x(1)\|^2_{C^TQC}$ has the effect of forcing the minimum value of the future

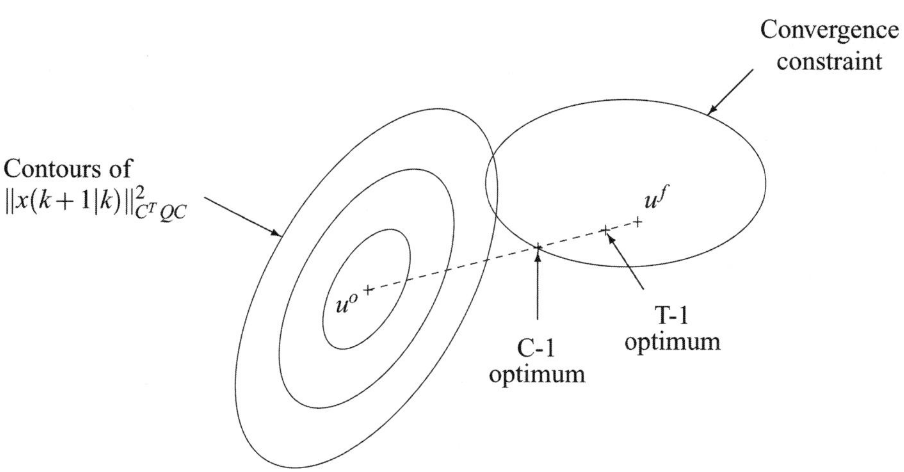

Figure 6.1 The optimisation of Algorithms C-1 and T-1 in the space of u(kjk) for the case that input and terminal state constraints are inactive

cost $\sum_{k=2}^{\infty}\|x(k)\|^2_{C^TQC}$ to be greater than that achievable with a less aggressive initial control move. Therefore the reduction in the stage cost $\|x(k+1|k)\|^2_{C^TQC}$ achieved by Algorithm C-1 for given $x(k)$ does not imply that this controller gives better performance than Algorithm T-1 with respect to the infinite horizon cost J_∞ in closed-loop operation. This section gives a comparison of the closed-loop optimality properties of the two algorithms by constructing infinite horizon costs of the form

$$\bar{J}_\infty = \sum_{k=0}^{\infty} l(x(k), \alpha(k)) \tag{21}$$

which are minimised along closed-loop trajectories of the plant under Algorithms T-1 and C-1, where $l(x, \alpha^*) \geq 0$ is satisfied by the solutions of (15) and (17) for $x(k) = x$.

To simplify notation we rewrite the prediction equation (5) as

$$x(k+1|k) = \phi(x(k)) + \psi(x(k))\alpha(k) \tag{22}$$

where $\phi(x) = f(x) + g(x)u^o(x)$ and $\psi(x) = g(x)\big(u^f(x) - u^o(x)\big)$. These dynamics are affine in the decision variable $\alpha(k)$ and it is therefore straightforward to adapt the inverse optimality approach of Reference 17 to the discrete-time case and thus derive a cost of the form (21) for which Algorithm T-1 is optimal if the terminal constraint (15c) is inactive for all $k \geq 0$.

Theorem 2: Let

$$\begin{aligned} l(x,\alpha) = {} & \|\phi(x) + \psi(x)\alpha\|^2_{C^TQC} - \|\phi(x) + \psi(x)\alpha^*(x)\|^2_{C^TQC} \\ & + \|x\|^2_P - \|\phi(x) + \psi(x)\alpha^*(x)\|^2_P \end{aligned} \tag{23}$$

where $\alpha^*(x)$ is the solution of (15) for $x(k) = x$, and assume that constraint (15c) is inactive for all $k \geq 0$. Then the control law of Algorithm T-1 minimises the cost of (21) over all $\alpha(k)$ $k = 0, 1, \ldots,$ satisfying input constraints (15b). Furthermore the optimal value of (21) is then $\|x(0)\|^2_P$.

Proof. Denote as $\bar{J}^*_\infty(x)$ the optimal value of (21) with $x(0) = x$. To prove that Algorithm T-1 is optimal for (21) with $l(x,\alpha)$ given by (23), we show that the HJB equation

$$\begin{aligned} & \min_\alpha \big\{ l(x,\alpha) + \bar{J}^*_\infty(\phi(x) + \psi(x)\alpha) \\ & \text{subject to} \quad \big|u^o(x) + \big(u^f(x) - u^o(x)\big)\alpha\big| \leq \bar{u} \big\} = \bar{J}^*_\infty(x) \end{aligned} \tag{24}$$

is satisfied with $\bar{J}^*_\infty(x) = \|x\|^2_P$ and that the minimising argument of the LHS is $\alpha^*(x)$.

With $\bar{J}^*_\infty(x) = \|x\|^2_P$, the solution of the minimisation in (24) is the value of α that minimises $\|\phi(x) + \psi(x)\alpha\|^2_{C^TQC+P}$ subject to input constraints, and this is given by

$\alpha^*(x)$ due to the equivalence of this problem with (15) whenever (15c) is inactive. Setting $\alpha = \alpha^*(x)$ in (23) and $\bar{J}^*_\infty(\phi(x) + \psi(x)\alpha) = \|\phi(x) + \psi(x)\alpha^*(x)\|^2_P$, the LHS of (24) is therefore given by

$$l(x, \alpha^*(x)) + \|\phi(x) + \psi(x)\alpha^*(x)\|^2_P = \|x\|^2_P$$

which completes the proof.

Remark 4: *Note that the monotonicity property (18) ensures that $l(x, \alpha)$ given by (23) is nonnegative.*

Remark 5: *With $l(x, \alpha)$ defined in (23) we have*

$$\bar{J}_\infty - J_\infty = \sum_{k=0}^{\infty} \|x(k)\|^2_P - \|\phi(x(k)) + \psi(x(k))\alpha^*(x(k))\|^2_{P+C^TQC}$$

along closed-loop trajectories under Algorithm T-1. It follows that Algorithm T-1 is optimal with respect to J_∞ if $\|x(k+1|k)\|^2_P - \|x(k)\|^2_P = -\|x(k+1|k)\|^2_{C^TQC}$ is satisfied for all $k \geq 0$.

Next consider the closed-loop properties of Algorithm C-1. From the optimisation (17) and prediction dynamics (22) it can be seen that Algorithm C-1 is a form of pointwise min-norm control law. Algorithm C-1 can therefore also be expected to be optimal with respect to a meaningful infinite-horizon cost [18]. Receding horizon control laws based on pointwise min-norm controllers are discussed in Reference 13, but this work uses a convergence constraint which is constructed with the explicit purpose of recovering the optimal controller with respect to a prespecified infinite-horizon cost whenever the value function in the convergence constraint coincides with the optimal cost function. On the other hand, the convergence constraint in Algorithm C-1 is chosen simply to ensure satisfaction of the monotonicity property (18) obtained under Algorithm T-1. It is possible, however, to extend the inverse optimality approach described above to determine whether a cost of the form (21) for Algorithm C-1 is optimal, as the following theorem shows.

Theorem 3: Let

$$\begin{aligned} l(x, \alpha) = & \frac{\|\phi(x) + \psi(x)\alpha\|^2_{C^TQC} - \|\phi(x) + \psi(x)\alpha^*(x)\|^2_{C^TQC}}{\mu^*(x)} \\ & + \|x\|^2_P - \|\phi(x) + \psi(x)\alpha^*(x)\|^2_P \end{aligned} \qquad (25)$$

where $\alpha^*(x)$, $\mu^*(x)$ are the solutions of (17) and (20) for $x(k) = x$, i.e. $\alpha^*(x) = \alpha_{\min}(\mu^*(x))$ and assume that constraint (15c) is inactive for all $k \geq 0$. Then the control law of Algorithm C-1 minimises the cost of (21) over all $\alpha(k)$ $k = 0, 1, \ldots$

satisfying input constraints (17b). Furthermore the optimal value of (21) is then $\|x(0)\|_P^2$.

Proof: Proceeding as in the proof of Theorem 2, when $l(x, \alpha)$ is given by (25) and $\bar{J}_\infty^*(x) = \|x\|_P^2$, the minimising value of α in (24) is also the minimising argument of

$$\min_\alpha \Big\{ \|\phi(x) + \psi(x)\alpha\|_{C^TQC}^2 + \mu^*(x)\|\phi(x) + \psi(x)\alpha\|_P^2$$

$$\text{subject to } \left|u^o(x) + \left(u^f(x) - u^o(x)\right)\alpha\right| \leq \bar{u}\Big\}$$

which from Theorem 1 is the solution $\alpha^*(x)$ of (17) whenever the terminal constraint (15c) is inactive. Furthermore the HJB equation (24) is satisfied along closed-loop trajectories under Algorithm C-1, since with $\alpha = \alpha^*(x)$ and $\bar{J}_\infty^*(\phi(x) + \psi(x)\alpha) = \|\phi(x) + \psi(x)\alpha^*(x)\|_P^2$, the LHS of (24) reduces to $l(x, \alpha^*(x)) + \bar{J}_\infty^*(\phi(x) + \psi(x)\alpha^*(x)) = \|x\|_P^2$.

Remark 6: *Note that $l(x(k), \alpha(k))$ in (25) is nonnegative for any $\alpha(k)$ satisfying the input and convergence constraints (17b–c) since $\|x\|_P^2 - \|\phi(x(k)) + \psi(x(k))\alpha^*(x(k))\|_P^2 \geq 0$ and $\alpha = \alpha^*(x(k))$ minimises $\|\phi(x(k)) + \psi(x(k))\alpha\|_{C^TQC}^2$ subject to (17b–c).*

Remark 7: *From the definition of μ^* in Theorem 1 it follows that $\mu^* \in [0, 1]$, and the convergence constraint (17c) is satisfied with equality whenever $\mu^* > 0$. Therefore $l(x, \alpha)$ as defined in (25) satisfies*

$$\lim_{\mu^*(x)\to 1} l(x, \alpha) = \|\phi(x) + \psi(x)\alpha\|_{C^TQC}^2$$

which implies that Algorithm C-1 is optimal with respect to J_∞ in the limit as $\mu^*(x(k)) \to 1$ if the terminal constraint (17b) is inactive for all $k \geq 0$. On the other hand, if both input constraints (17b) and the terminal constraint (17d) are inactive, then $\alpha_{\min}(\mu^*) \to 0$ as $\mu^* \to 0$, and (25) gives

$$\lim_{\mu^*(x)\to 0} l(x, \alpha) = \frac{\|\phi(x) + \psi(x)\alpha\|_{C^TQC}^2}{\mu^*(x)}$$

indicating that Algorithm C-1 minimises J_∞ as $\mu^* \to 0$, if (17b,d) are inactive for all $k \geq 0$.

Comparison of (23) and (25) shows that the stage cost $\|x(k+1|k)\|_{C^TQC}^2$ of J_∞ is penalised more heavily in the cost minimised by Algorithm C-1 than in the cost for which Algorithm T-1 is optimal. This implies that Algorithm C-1 is likely to give better performance with respect to J_∞ than Algorithm T-1. Remarks 5 and 6 support this conclusion since both Algorithms T-1 and C-1 tend towards the optimal control with respect to J_∞ as the monotonicity property (18) becomes more stringent, whereas Algorithm C-1 can also recover optimal performance as μ^* tends to zero, which corresponds to the convergence constraint becoming less stringent.

6.5 Simulation example

The plant model is given by the second order single-input bilinear system:

$$x(k+1) = Ax(k) + (B + Fx(k))u(k) \tag{26a}$$

$$A = \begin{bmatrix} 0.28 & -0.78 \\ -0.78 & -0.59 \end{bmatrix} \quad B = \begin{bmatrix} 0.71 \\ 1.62 \end{bmatrix} \quad F = \begin{bmatrix} 0.34 & 0.36 \\ 0.41 & -0.65 \end{bmatrix} \tag{26b}$$

with input constraint

$$|u(k)| \leq 0.5, \ k = 0, 1, \ldots$$

The infinite horizon performance index is given by (3) with

$$C = \begin{bmatrix} -0.69 & 0.20 \end{bmatrix} Q = 1$$

Due to nonminimum-phase characteristics, the origin $x = 0$ of (26) is unstable under the control law $u^o(x)$ defined by (7).

Figure 6.2 shows the infinite horizon costs J_∞ corresponding to closed-loop responses of (26) under the control laws of Algorithms C-1 and T-1. These were simulated for 100 initial conditions equispaced around the unit circle in state space. It can be seen from this figure that Algorithm C-1 achieves a lower cost than Algorithm T-1 for every initial condition on the set considered.

The better performance of Algorithm C-1 can be explained by considering the closed-loop responses for a particular initial condition shown in Figures 6.3–6.6. The interpolation variable $\alpha(k)$ for Algorithm C-1 is smaller than that for Algorithm T-1 for all $k > 0$ (Figure 6.3), indicating that Algorithm C-1 follows more closely the unconstrained optimal control law u^o. This is reflected in the input response of Algorithm C-1 (Figure 6.4), which is initially more aggressive than that of Algorithm T-1. In fact the interpolation variable for Algorithm C-1 converges to zero in the steady state even though u^o does not stabilise the equilibrium of the closed-loop system at $x = 0$ due to the unstable zero dynamics of the model. The reason for this apparent contradiction is that Algorithm C-1 does not drive $x(k)$ to the origin, as can be seen from the nonzero steady-state value of $\|x(k)\|_P^2$ in Figure 6.5, but instead converges to an equilibrium which is stable under u^o and for which $Cx = 0$. On the other hand, Algorithm T-1, which places more emphasis on minimising $\|x(k)\|_P^2$, forces $x(k)$ to converge to zero and therefore cannot benefit from a value of $\alpha = 0$ in the steady state.

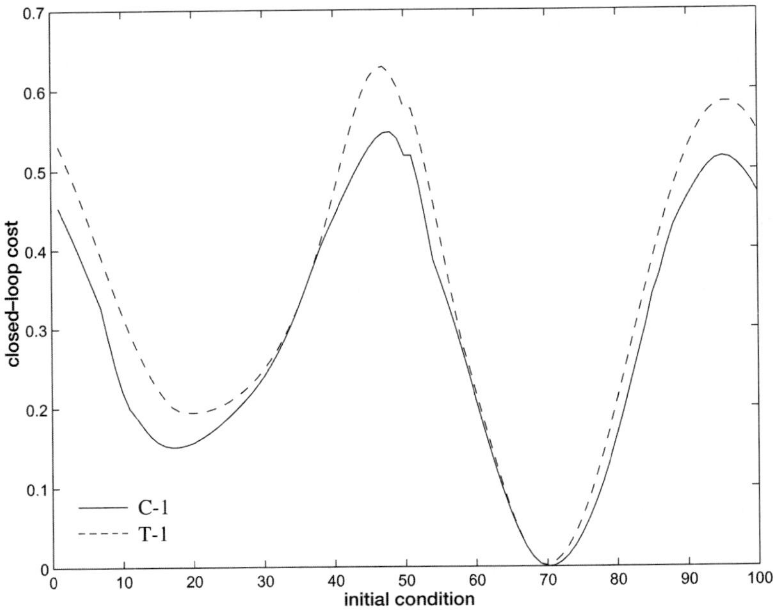

Figure 6.2 *Comparison of the closed-loop costs J_∞ obtained under Algorithm C-1 and T-1 for 100 initial conditions $x(0)$ equally spaced around the unit circle*

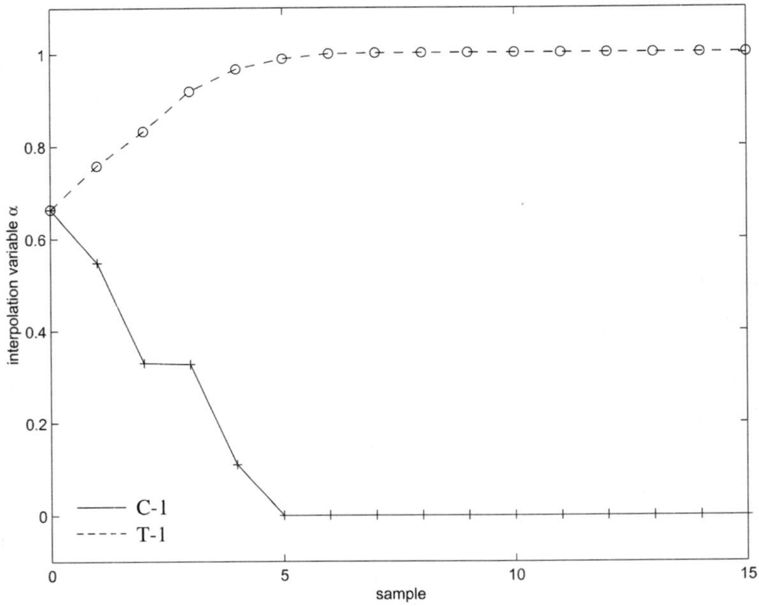

Figure 6.3 *Time-histories of the interpolation variable $\alpha(k)$ for Algorithms C-1 and T-1 with initial condition 95*

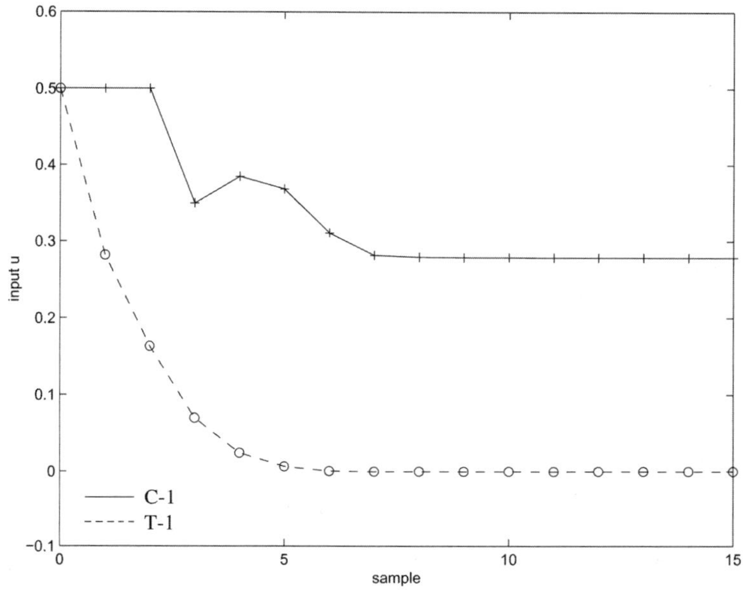

Figure 6.4 Input responses of Algorithms C-1 and T-1 with initial condition 95

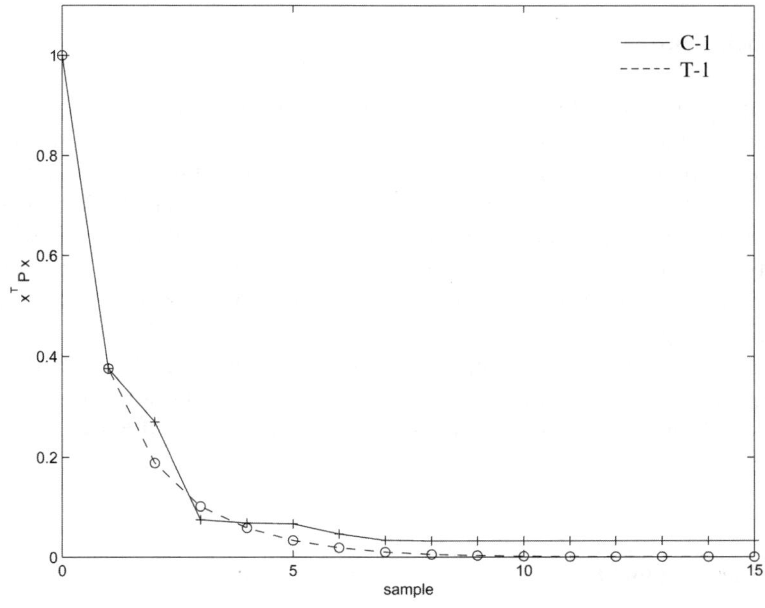

Figure 6.5 Time-histories of the value function $\|x(k)\|_P^2$ for Algorithms C-1 and T-1 with initial condition 95

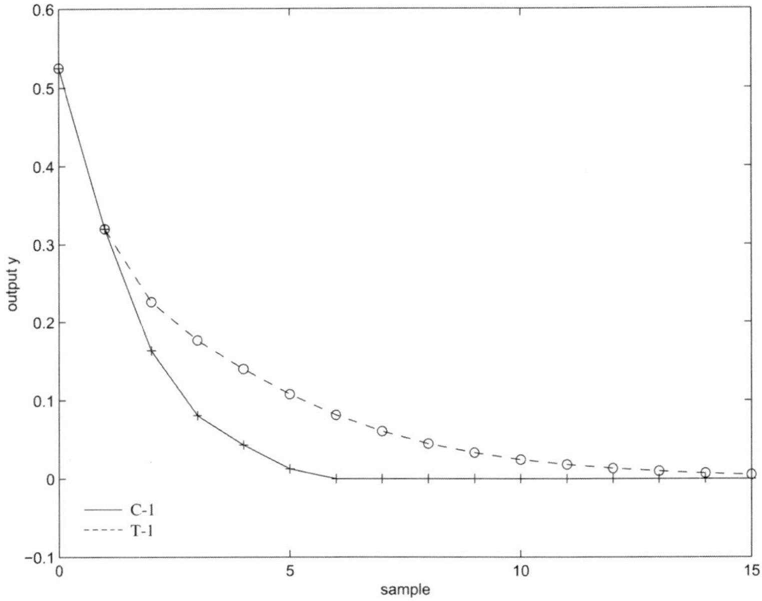

Figure 6.6 Output responses $y(k) = Cx(k)$ of Algorithms C-1 and T-1 with initial condition 95

6.6 Conclusions

This chapter describes MPC algorithms for a class of input affine nonlinear systems that are subject to input and/or state constraints. Predicted future control trajectories are obtained by interpolating between an unconstrained optimal control law and a control law which is locally stabilising for the constrained plant. This approach results in a univariate online optimisation, the computational burden of which is minimal. We compare two methods of ensuring closed-loop stability: addition of a suitable terminal weight in the receding horizon objective versus inclusion of an explicit convergence constraint on closed-loop trajectories. The latter method necessarily achieves a smaller open-loop predicted cost, and by exploiting the relationship between the two techniques, we show that the latter also achieves better closed-loop performance under a range of operating conditions.

6.7 Acknowledgment

The authors thank the Engineering and Physical Sciences Research Council for financial support.

6.8 References

1 ZHENG, A.: 'A computationally efficient nonlinear MPC algorithm'. *Proc. ACC*, Albuquerque, 1997, pp. 1623–7

2 ZHENG, A.: 'Some practical issues and possible solutions for nonlinear predictive control', in 'Nonlinear model predictive control: assessment and future directions for research' (Birkhäuser, 1999) pp. 129–44

3 KOUVARITAKIS, B., ROSSITER, J.A., and CANNON, M.: 'Linear quadratic feasible predictive control', *Automatica*, 1998, **34** (12), pp. 1583–92

4 KOUVARITAKIS, B., CANNON, M., and ROSSITER, J.A.: 'Stability, feasibility, optimality and the degrees of freedom in constrained predictive control', in 'Nonlinear model predictive control: assessment and future directions for research' (Birkhäuser, Basel, 1999), pp. 99–113

5 BLOEMEN, H.H.J., CANNON, M., and KOUVARITAKIS, B.: 'Interpolation in MPC for discrete-time bilinear systems'. ACC 2001, to be published

6 SZNAIER, M., and DAMBORG, M.J.: 'Suboptimal control of linear systems with state and control inequality constraints'. IEEE CDC, 1987, pp. 761–2

7 DE NICOLAO, G., MAGNI, L., and SCATTOLINI, R.: 'Stabilising receding-horizon control of nonlinear time-varying systems', *IEEE Trans.*, 1998, **AC-43** (7), pp. 1030–6

8 CHEN, H., and ALLGÖWER, F.: 'A quasi-infinite horizon nonlinear model predictive control scheme with guaranteed stability', *Automatica*, 1998, **14** (10), pp. 1205–17

9 MAYNE, D.Q., RAWLINGS, J.B., RAO, C.V., and SCOKAERT, P.O.M.: 'Constrained model predictive control: stability and optimality', *Automatica*, 2000, **36**, pp. 789–814

10 SZNAIER, M., and DAMBORG, M.J.: 'Heuristically enhanced feedback control of constrained discrete time linear systems', *Automatica*, 1990, **26**, pp. 521–32

11 BEMPORAD, A.: 'A predictive controller with artificial Lyapunov function for linear systems with input/state constraints', *Automatica*, 1998, **34** (10), pp. 1255–60

12 FREEMAN, R.A.: 'An adaptive approach to Lyapunov design in nonlinear optimal stabilisation problems'. IEEE CDC, San Diego, 1997, pp. 2339–44

13 PRIMBS, J., NEVISTIC, V., and DOYLE, J.C.: 'On receding horizon extensions and control Lyapunov functions'. Proc. ACC, Philadelphia, 1998, pp. 3276–80

14 NIJMEIJER, H., and VAN DER SCHAFT, A.J.: 'Nonlinear dynamical control systems' (Springer-Verlag, New York, 1990)

15 KOTHARE, M.V., BALAKRISHNAN, V., and MORARI., M.: 'Robust constrained model predictive control using linear matrix inequalities', *Automatica*, 1996, **32** (10), pp. 1361–79

16 MICHALSKA, H., and MAYNE, D.Q.: 'Robust receding horizon control of constrained nonlinear systems', *IEEE Trans.*, 1993, **AC-38**, pp. 1623–32

17 MAGNI, L., and SEPULCHRE, R.: 'Stability margins of nonlinear receding-horizon control via inverse optimality', *Syst. Control Lett.*, 1997, **32**, pp. 241–5

18 FREEMAN, R.A., and KOKOTOVIC, P.V.: 'Robust nonlinear control design' (Birkhäuser, Boston, 1996)

Part III

Chapter 7

Closed-loop predictions in model based predictive control of linear and nonlinear systems

B. Kouvaritakis, J.A. Rossiter and M. Cannon

Abstract

Conventional model based predictive control (MPC) algorithms consider predicted control moves to be the degrees of freedom over which the predicted performance cost is to be optimised. Recent work introduced a closed-loop prediction paradigm which enables the overall MPC strategy to be decomposed into the design of a fixed-term linear inner loop with a variable nonlinear loop wrapped around it. By effecting the design of the innerloop off-line, this paradigm affords significant advantages for both the control of uncertain linear and of nonlinear systems. This chapter exploits these advantages to develop effective predictive control algorithms which have guaranteed closed-loop stability and asymptotically can yield optimal dynamic behaviour.

7.1 Introduction

Model based predictive control (MPC) has been an active area of research, starting with a proliferation of heuristic algorithms which were unified and extended by generalised predictive control (GPC) [1]. Despite their popularity, these early developments lacked a general guarantee of closed-loop stability. Equality endpoint constraints [2, 3] offer a convenient means of overcoming this problem [4–6]. Like most other MPC literature, References 4 and 5 used an open-loop prediction formulation in which predicted control moves are considered to be the degrees of freedom. In contrast to this, Reference 6 used a prestabilised loop and the input c to this loop was adopted as the decision variable. This established a new paradigm in MPC which offers some very distinct advantages (discussed below) which are to be exploited in this chapter.

Equality endpoint constraints exact highly tuned predicted behaviour from the system dynamics and thus can come into conflict with constraints due to physical limits (e.g. input saturation or slew rates, output safety limits, etc.). For this reason the use of alternative endpoint constraints was explored. For example, a stable MPC algorithm was derived by constraining the model state-vector to reach a stable subspace within a given control horizon [7]. However, this approach was based on open-loop predictions and considered future control moves to be the degrees of freedom. Thus, whereas it was possible to relax the equality endpoint constraint in respect of outputs, it was still necessary to retain input endpoint constraints, in that nonzero future control moves had to be taken to be finite in number. On account of this, infeasibility (i.e. conflict with physical limits) could still be an issue. A possible avenue for the complete removal of equality endpoint constraints was provided by a characterisation of all stabilising predictions [8]. This approach [9, 10], like that of Reference 6, uses closed-loop predictions, and can be shown to be closely related to other alternatives [11–13] which are cast in a dual-mode open-loop prediction framework. Dual-mode strategies have been used effectively in linear MPC but have also provided useful solutions in nonlinear MPC [14, 15].

The aim of the current chapter is to emphasise the importance of the closed-loop prediction paradigm introduced in Reference 6 and in particular to show its strength in handling two important problems: (i) MPC of linear but uncertain systems; (ii) nonlinear MPC. Over and above its numerical advantages [9, 16] and benefits derived from the expansion of the prediction class (through the elimination of input endpoint constraints), it provides a significant simplification both with respect to constraint handling and optimisation. Thus, in the case of linear uncertain systems, it allows for a systematic way to separate the issue of robustness from that of feasibility. In particular, under the assumption that c will be deployed to achieve feasibility, the prestabilised loop can be designed as per unconstrained MPC to yield a linear fixed term controller which optimises nominal performance and robustness [17]. The full robust constrained MPC problem then becomes that of simply finding the minimum norm c which achieves feasibility over the entire uncertainty class. Techniques developed recently [18] provide an effective means for attaining robust feasibility for a wide class of initial conditions at a trivial on-line computational cost. The approach is based on ellipsoidal invariant sets defined on an augmented autonomous prediction model. By its very definition, the prediction class contains the extension to current time of the previously computed input trajectory; for convenience we shall refer to this extension as the ‘tail’. The inclusion of the tail in the prediction class ensures that: (i) feasibility at the current instant guarantees feasibility at the next; (ii) the minimised norm of the current sequence of c values is monotonically decreasing. These two properties in combination with the robust stability/optimality of the prestabilised loop establish the overall closed-loop stability and asymptotic optimality of the proposed algorithm.

In the context of nonlinear control, the closed-loop prediction paradigm proves just as useful. The ideas are similar in that a fixed-term controller which, in the absence of constraints, is stabilising and optimal (in an asymptotic sense) is used to generate the closed-loop predictions. This, once again, converts the MPC problem to one of finding the minimal norm c which achieves feasibility. We tackle this issue here using a mild Lipschitz condition in conjunction with polyhedral invariant sets to define a prediction class which achieves feasibility over an infinite horizon with a guarantee of feasibility at the next sampling instant. Lipschitz conditions have been used in a similar context before [14, 15]; however, here they are applied at current time (not at the end of a finite control horizon) and are applied to a driven system (not a free responding state feedback system). In addition, as mentioned above we achieve feasibility and therefore stability/optimality through the use of polyhedral sets and the concept of the tail.

7.2 Review of earlier work

The closed-loop prediction paradigm was first introduced in the context of stable generalised predictive control (SGPC) [6] and was later also adopted by algorithms such as the Reference Governor [21, 22] to achieve setpoint conditioning [23, 24]. As explained in the introduction, SGPC used a prestabilised loop of the form shown in Figure 7.1, where $a(z)$, $b(z)$ are two relatively prime polynomials in z^{-1} of degree $n, n-1$, respectively, $\Delta(z) = 1 - z^{-1}$, and Δu, y denote control increment and output variables; the $X(z)$, $Y(z)$ polynomials are chosen so as to satisfy the Bezout identity,

$$b(z)Y(z) + a(z)\Delta(z)X(z) = 1 \tag{1}$$

Using (1) it is easy to show that both transfer functions from c to the input and to the output have 1 as their denominator. Hence a sequence of c values which reaches a steady-state value, say c_∞, within a finite number of steps, say v, will cause both control increments and the output values to reach a steady state within a finite number of steps. Furthermore the steady-state value for the control increments is 0, whereas that for the output can be made to be a prespecified value r (for

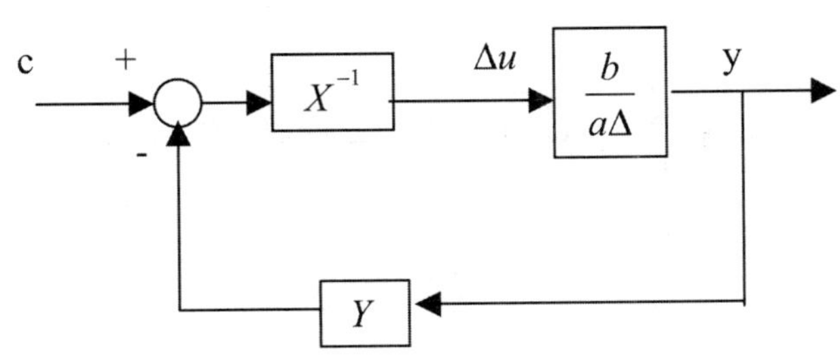

Figure 7.1 SGPC closed-loop prediction system

convenience we consider constant setpoints r only) providing that c_∞ is chosen appropriately.

Thus the prediction system of Figure 7.1 automatically satisfies the endpoint equality constraints that enable one [6] to write the usual infinite horizon MPC cost as a finite horizon one and in particular as a quadratic function of c, the vector comprising the v values of the sequence c (excepting c_∞):

$$J = \sum_{t=1}^{\infty} (r - y_{t+1})^2 + \lambda(\Delta u_t)^2 = \underset{\rightarrow}{c}^T S \underset{\rightarrow}{c} + \beta^T \underset{\rightarrow}{c} + \gamma \tag{2}$$

In the absence of physical constraints, minimisation of this cost over $\underset{\rightarrow}{c}$ results in a feedback loop with a linear fixed-term controller (Figure 7.2), where the polynomials M, N satisfy the Bezout identity

$$b(z)N(z) + \Delta a(z)M(z) = p(z) \tag{3}$$

with $p(z)$ being the closed-loop pole polynomial resulting from the process of the receding horizon application of SGPC, i.e. from the minimisation of J at each time instant and implementation of the current optimal control increment. Now (3) does not have a unique solution for $M(z)$, $N(z)$; indeed, given a particular solution pair $M_o(z)$, $N_o(z)$ we can write the general class of solutions as

$$M(z) = M_o(z) + b(z)Q(z), \quad N(z) = N_o(z) - a(z)Q(z) \tag{4}$$

Thus, without affecting the dynamic performance achieved through the minimisation of J, we can replace the $M(z)$, $N(z)$ in Figure 7.2 by those of (4) to get Figure 7.3. $Q(z)$, which must be a polynomial or stable transfer function, can therefore be treated as a Youla parameter, available for optimising robustness properties (e.g. tolerance to model uncertainty or insensitivity to disturbances) without affecting the dynamic performance from setpoint r to control increments or from r to the system output.

In the presence of physical limits (e.g. input saturation, or safety limits on system states) in conventional MPC it is not possible to derive a fixed-term linear feedback implementation of SGPC. The reason for this is that minimising J subject to constraints leads to a nonlinear feedback law. However, due to the use of the closed-loop prediction paradigm of SGPC, namely due to the fact that the decision

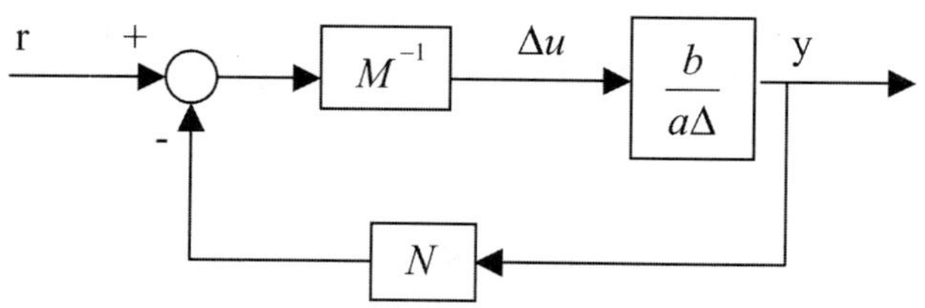

Figure 7.2 The SGPC fixed-term feedback control loop

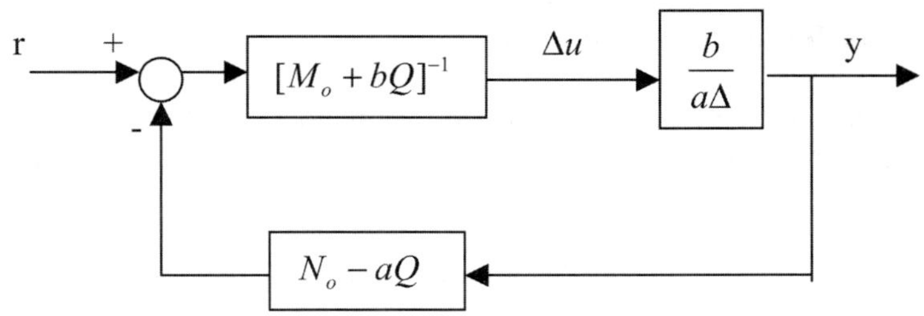

Figure 7.3 SGPC feedback control loop with Youla parameter

variable used in the minimisation is c, not the control increments adopted by other MPC strategies, it is still possible to separate the overall feedback implementation of SGPC into an inner linear fixed-term loop with a variable nonlinear loop around it (see Figure 7.4).

This figure exposes one of the major strengths of the closed-loop paradigm adopted by SGPC over the usual paradigm adopted by other MPC algorithms. The implementation of those other algorithms involves one variable/nonlinear feedback loop whereas, as explained above, with the closed-loop paradigm we are able to separate the control law into an inner feedback law which is linear and fixed-term and an outer feedback law which is variable and nonlinear. In the sequel we shall exploit this separation in order to address the problems of: (i) MPC of linear uncertain systems; (ii) nonlinear MPC. It is pointed out, of course, that implicit in the above figure is the assumption that, at each time instant, a feasible sequence c can be found such that the predicted behaviour does not violate any of the physical constraints. As will be seen below, for a particular choice of $M_o(z), N_o(z)$ feasibility rather than constrained optimisation is the challenging issue to address.

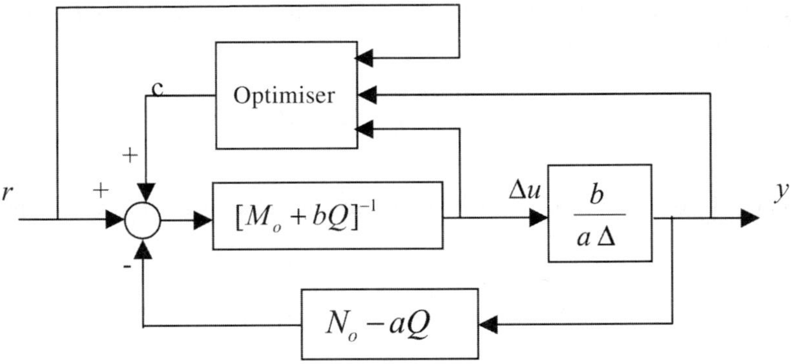

Figure 7.4 Constrained SGPC closed-loop paradigm

7.3 MPC for linear uncertain systems

For simplicity in the sequel we shall consider regulation only ($r = 0$) and shall examine the class of linear uncertain systems for which the error system has dynamics:

$$x_{k+1} = \tilde{A}x_k + \tilde{B}\Delta u_k, \quad e_k = Cx_k \tag{5}$$

where

$$\tilde{A} = A + \delta A, \quad \tilde{B} = B + \delta B \tag{6}$$

The uncertainty class of (6) is assumed to be polytopic with extreme points (A_i, B_i) so that it can be described alternatively by the set

$$\Omega = \{(\tilde{A}, \tilde{B}) : (\tilde{A}, \tilde{B}) = \sum_{i=1}^{n_p} \eta_i (A_i, B_i), \ \eta_i \geq 0, \ \sum_{i=1}^{n_p} \eta_i = 1\} \tag{7}$$

The matrices A, B can be considered to give a nominal description with transfer function

$$G(z) = \frac{b(z)}{\Delta a(z)} \tag{8}$$

For convenience, let us ignore uncertainty for the moment. Under the assumption that c takes care of feasibility, the inner loop of Figure 7.4 remains within physical limits and can thus be considered to be unconstrained. Thus it is no longer necessary to restrict the choice of $M_o(z), N_o(z)$ to those which minimise J under the SGPC endpoint equality constraints. Indeed there is no reason why these polynomials could not, for example, be chosen so as to generate the unconstrained LQ optimal control for the model of (8).

Lemma 1 [9]: Let $M(z) = M_o(z), N(z) = N_o(z)$ be such that the feedback system of Figure 7.5 for $c = 0$ and in the absence of constraints is LQ optimal with respect to the infinite horizon cost

$$J = \sum_{k=0}^{\infty} e_{k+1}^2 + \lambda \Delta u_k^2 \tag{9}$$

Let also C denote the set of c which are feasible when constraints are present. Then

$$\arg\left\{ \min_{c \in C} J \right\} = \arg\left\{ \min_{c \in C} J_2 \right\} \ J_2 = \sum_{k=0}^{\nu-1} c_k^2 \tag{10}$$

Proof: The dependence of J on c must be such that in the case of inactive constraints the minimising solution is zero, namely $J = \underset{\rightarrow}{c}^T S \underset{\rightarrow}{c} + \gamma$ for S symmetric

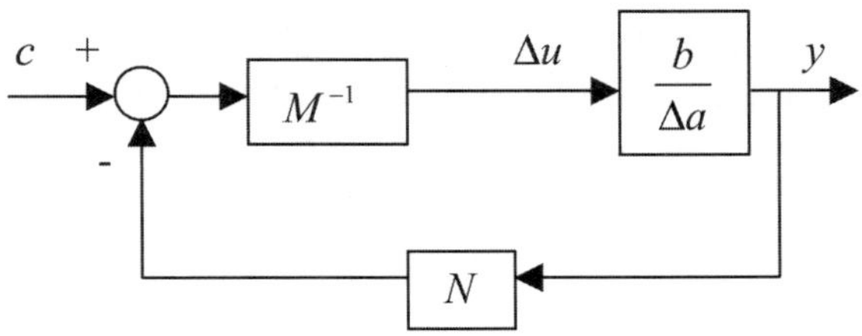

Figure 7.5 The closed-loop paradigm for the regulation problem

positive definite and γ a positive constant. Detailed calculation of the quadratic term [9] of the cost shows that $S = I$.

The lemma above makes practical sense because, given the (unconstrained) optimality of the feedback loop, c ideally should be zero but is needed to cater for feasibility. For constrained optimality therefore we should be looking for the smallest c which achieves feasibility.

The arguments above apply equally to the uncertain case. Thus if the nominal description A, B is considered to be the most representative member of the class, then $M_o(z)$, $N_o(z)$ could be chosen to give unconstrained LQ optimality for the nominal and the Youla parameter could be chosen to guarantee robust stability over the uncertainty class. Alternatively $M(z)$, $N(z)$ could be chosen to give worst case optimality over the class. The particular choice of these matrices is not important so long as it can be claimed that they bestow the unconstrained linear feedback system of Figure 7.5 with some optimality property in addition, of course, to robust stability. Then, clearly, in the constrained case nonzero c will be needed to cater for feasibility but should be as small as possible, and indeed should return to zero just as soon as constraints become inactive. Thus the cost J_2 is just as valid for the uncertain case as it is for the case with no uncertainty. This observation shows that the closed-loop paradigm produces a significant simplification of the MPC problem because it delegates a lot of the optimisation burden onto the choice of M, N which can be performed off-line, whereas it reduces on-line optimisation to a feasibility problem; find the smallest feasible c. This is in direct contrast to other MPC strategies [25] which perform worst-case optimisation on-line.

Lemma 2: Under the assumption that M, N robustly stabilise the unconstrained feedback system of Figure 7.5, the constrained system of the figure is also robustly stable for all feasible c which converge to zero.

Proof: This is rather obvious given the robust stabilising property of M, N.

This lemma therefore reiterates the point made earlier that the key issue is finding feasible c. A convenient way of tackling this problem is to restrict the choice of c in such a way that feasibility at current time implies feasibility at the next. The benefits of this approach are stated in the following result.

Theorem 1: Given that M, N are robustly stabilising and providing that c is chosen so that it not only is feasible over the uncertainty class at the current time but also ensures the existence of feasible c at the next, then the system of Figure 7.5 is stable and will converge to the optimal solution just as soon as it becomes feasible.

Proof: This is based on the fact that under the assumption of the theorem the same sequence c (except for its first element) can be used at the next time instant. Dropping the first element of c, however, incurs a reduction in J_2 which in fact could possibly be made smaller still upon optimisation. On account of this monotonic decrease of cost, the sequence c will converge to zero and thus by Lemma 2 will yield closed-loop stability. It also ensures a return to the unconstrained optimality of the system of Figure 7.5 just as soon as this is possible, namely when $c = 0$ is feasible.

Thus the only remaining problem is the attainment of a recursive guarantee of feasibility. This can be achieved through a recent technique [18] that enlarges the set of initial conditions under which the state trajectory is guaranteed to stay within an invariant ellipsoidal set and which at the same time is also feasible over the whole uncertainty class. Thus let the state space description of the system of Figure 7.5 be:

$$x_{k+1} = \tilde{\Phi} x_k + \tilde{B} c_k \quad e_k = C x_k \quad \tilde{\Phi} = \tilde{A} - \tilde{B} K \tag{11}$$

Note that if M, N are chosen so as to give unconstrained LQ optimality for the nominal model of (8) then the $\tilde{A}, \tilde{B}$ matrices of (11) will be the same as those of (5) and K can be taken as the optimal state feedback controller derived from the appropriate steady-state Riccati equation. However, in the case of general M, N (11) can be derived simply through a nonminimal state-space realisation of the description of the system of Figure 7.5. As such, the state-space matrices $\tilde{A}, \tilde{B}$ in (11) will be of higher dimension than the order of (8) but the sets of models implied by (11) will be related in an affine manner to the sets of models in (5), thereby making it possible to transform linearly the polytopic class of (5)–(7) to get an equivalent polytopic class for (11). For simplicity here we shall assume that the M, N of Figure 7.5 are such that the system can be expressed directly (without recourse to nonminimal realisations) in terms of the state feedback state-space equations of (11). Then prediction trajectories for the system of Figure 7.5 can be generated from the response of the autonomous system:

$$z_{k+1} = \tilde{\Psi} z_k, \quad z_k = \begin{bmatrix} x_k \\ \underset{\rightarrow}{c_k} \end{bmatrix}, \quad \tilde{\Psi} = \begin{bmatrix} \tilde{A} - \tilde{B}K & \tilde{B}E \\ 0 & T \end{bmatrix}, \quad E = [I \quad 0 \quad \ldots \quad 0] \tag{12}$$

where T is the zero block matrix with Is in its block superdiagonal. Using this formulation it is easy to show that for P positive definite, the ellipsoid

$$E_z = \{z : z^T P^{-1} z \leq 1\} \tag{13}$$

is invariant [25], namely that $z_{k+1} \in E_z$ given that $z_k \in E_z$ if

$$\begin{bmatrix} P & P\tilde{\Psi}^T \\ \tilde{\Psi}P & P \end{bmatrix} \geq 0 \tag{14}$$

This condition ensures that, for an initial condition in E_z, z remains in E_z at all prediction times. Then, given (13) it is easy to ensure feasibility by invoking

$$\hat{K}^T P \hat{K} \leq \bar{u}^2 \tag{15}$$

Here for simplicity we only consider saturation limits of the form $-\bar{u} \leq u \leq \bar{u}$; other forms of linear constraints on inputs and states can be included easily. For the recursive guarantee of feasibility therefore (14)–(15) must hold true over the whole uncertainty class. However, both these conditions are linear matrix inequalities (LMI) and need only be invoked (off-line) at the corners (A_i, B_i) of the uncertainty class. Hence, providing that a positive definite P satisfying (14) and (15) exists, and that the initial state is such that a solution for $\underset{\rightarrow}{c}$ satisfying (13) exists, then the recursive guarantee of feasibility is ensured and with it we have the guarantee of stability and convergence to the optimal unconstrained control law, just as soon as this becomes feasible.

Algorithm 1:

(Off-line) Step 0: Compute P satisfying (14) and (15) for all corners of Ω.

(On-line) Step 1: At each instant of time compute the smallest (in the 2-norm sense) $\underset{\rightarrow}{c}$ *which satisfies (13). Use the first element of the optimal* $\underset{\rightarrow}{c}$ *in* $\Delta u = -Kx + \vec{c}$ *to compute the current optimal control increment and from this the optimal current input.*

(On-line) Step 2: Implement this optimal value and at the next sampling instant go back to Step 1.

Theorem 2: Under the assumption that K is robustly stabilising, and that the initial state is such that a feasible solution $\underset{\rightarrow}{c}$ to (13) exists, then Algorithm 1 has guaranteed closed-loop stability and will converge to the unconstrained optimum just as soon as this becomes feasible.

Proof: This has already been sketched above and of course presupposes the existence of a matrix positive symmetric P which satisfies (13) over the uncertainty class.

7.4 Invariance/feasibility for nonlinear systems

The fixed-term structure of the feedback loop of the closed-loop paradigm (Figure 7.5) strictly speaking is not suitable for use when the linear plant b/a is replaced by one governed by nonlinear dynamics. The reason for this is that optimality is then state-dependent. It is possible to relax the requirement for a fixed-term controller, but this will not be pursued further here. Instead it is argued that the main issue in nonlinear MPC is the computational burden which, due to the nonconvexity of both cost and feasible region, often makes the application of MPC impracticable. We are therefore looking for substantial reductions in the computation load, and the closed-loop paradigm of Figure 7.5 offers a convenient (albeit suboptimal) means of achieving this. The ideas parallel those given in Section 7.3 for the linear case. Thus we make the assumption that the initial condition is such that for a fixed M, N or K there exists $\underset{\rightarrow}{c}$ such that the state can be steered into a feasible invariant set; for simplicity here we shall avoid the use of nonminimal realisations by adopting the state feedback controller K. Then it is possible to guarantee recursively the existence of such $\underset{\rightarrow}{c}$ at all future times. Under the assumption that within the invariant set, for $\underset{\rightarrow}{c} = 0$, K yields a desirable closed-loop response, it is sensible to adopt J_2 as the cost to minimise at each sampling instant. Therefore, it is possible to invoke similar arguments to those used in Section 3 to prove that the cost will be a monotonically decreasing function of time thereby giving the guarantee of closed-loop stability and asymptotic convergence to the desirable (but no longer optimal) unconstrained behaviour of the state feedback control law $u = -Kx$. The details of this analysis are similar to those given for the linear case and will not be repeated here. What is different, however, is the treatment of invariance and feasibility; it turns out that it is no longer convenient to use ellipsoidal invariant sets which, therefore, in the sequel are replaced by polyhedral sets.

For simplicity we consider input constraints only and take these to be

$$-\bar{u} \leq u_k = -Kx_k + c_k \leq \bar{u} \tag{16}$$

while the closed-loop dynamics, given by

$$x_{k+1} = F(x_k, c_k), \quad F(0,0) = 0 \tag{17}$$

are assumed to be stable for $c_k = 0$ and for release from initial conditions in a given set X, containing the origin. Then the problem we are considering here becomes simply that of deriving conditions on c which ensure feasibility. We do this by first applying the state transformation:

$$x = W_o q, \; q = V_o x \qquad V_o = W_o^{-1} \tag{18}$$

according to which (17) becomes

$$q_{k+1} = f(q_k, c_k) \qquad f(0,0) = 0 \tag{19}$$

Then in place of (11) we write

$$z_{k+1} = \begin{bmatrix} f(q_k, E\underset{\rightarrow}{c_k}) \\ T\underset{\rightarrow}{c_k} \end{bmatrix} \quad z_i = \begin{bmatrix} q_i \\ \underset{\rightarrow}{c_i} \end{bmatrix} \tag{20}$$

and in place of the ellipsoid E_z of (13) we consider the polyhedral set [19, 20]:

$$P_z = \{z : |\hat{z}| \leq \bar{z}\} \quad \hat{z} = \begin{bmatrix} \hat{q} \\ \hat{\underset{\rightarrow}{c}} \end{bmatrix} = \hat{W}z \quad \bar{z} = \begin{bmatrix} \alpha \\ \gamma \end{bmatrix} \quad \hat{W} = \begin{bmatrix} W & Q \\ R & S \end{bmatrix} \tag{21}$$

where α, γ are two positive vectors (yet to be determined) of conformal dimensions to the partition of z; the matrices W, Q represent degrees of design freedom which can be used to advantage as discussed below. Note that henceforth absolute values together with the relevant inequalities are used on an element-by-element basis. The polyhedron P_z becomes invariant if

$$\left| \hat{W} \begin{bmatrix} f(q_k, E\underset{\rightarrow}{c_k}) \\ T\underset{\rightarrow}{c_k} \end{bmatrix} \right| < \bar{z} \quad \forall z \in P_z \tag{22}$$

However, from (21) we have

$$z = \hat{W}^{-1}\hat{z} = \hat{V}\hat{z} \tag{23}$$

so that a convenient way to invoke the invariance condition is to assume that for all $z \in P_z$

$$\left| \hat{W} \begin{bmatrix} f(q_k, E\underset{\rightarrow}{c_k}) \\ T\underset{\rightarrow}{c_k} \end{bmatrix} \right| \leq \Gamma \left[\left| \begin{matrix} \hat{q} \\ \underset{\rightarrow}{c} \end{matrix} \right| \right] \tag{24}$$

where Γ is a positive matrix. This is a perfectly sensible assumption to make; for example, for the case of linear systems with $Q = 0, R = 0, S = I$, (24) would obviously be true for

$$\Gamma_{11} = |W(A - BK)V|, \Gamma_{12} = |WB| \tag{25}$$

where Γ_{11}, Γ_{12} are conformal partitions of Γ. Furthermore if V is chosen to be the eigenvector matrix of $A - BK$, then Γ_{11} would be diagonal with diagonal elements equal to the moduli of the closed-loop eigenvalues, all of which moduli would be less than 1.

The combination of (21) and (24) gives the invariance condition:

$$\Gamma\bar{z} \leq \bar{z} \tag{26}$$

In deriving the above we have used the fact that invariance must hold true for all $z \in P_z$. Condition (26) ensures the invariance of P_z but we also need feasibility, which is catered for by requiring:

$$\begin{aligned} |u| = |-Kx + c| = \left|\hat{K}\hat{z}\right| \leq \bar{u}, \quad \forall z \in P_z \\ \hat{K} = [-K \quad E]\hat{V} \end{aligned} \tag{27}$$

To ensure that this holds true for all z in the invariant set therefore we must enforce the condition

$$|\hat{K}|\bar{z} \leq \bar{u} \tag{28}$$

The existence of $\bar{z}$ which satisfies both the invariance condition (26) and feasibility condition (28) guarantees the existence of a feasible and invariant set P_z.

Given the definition and conditions for the existence of invariant feasible sets we can apply the arguments of Section 3 to propose the following nonlinear MPC algorithm.

Algorithm 2:

(Off-line) Step 0: Choose W and compute the smallest Γ for which (26) and (28) hold true.

(On-line) Step 1: At each time instant minimise over $\underset{\rightarrow}{c}$ the cost J_2 subject to (21).

(On-line) Step 2: Use the first element of the optimal $\underset{\rightarrow}{c}$ to compute the current control move; implement and at the next time instant go to Step 1.

Remark 1: *Condition (21) is linear in $\underset{\rightarrow}{c}$ and thus Step 1 of the algorithm can be executed efficiently using quadratic programming.*

Remark 2: *From (21) it is clear that the algorithm can only be applied to the set of initial states:*

$$P_x(\alpha, \gamma) = \left\{ x : \left| \hat{W} \begin{bmatrix} V_o x \\ \underset{\rightarrow}{c} \end{bmatrix} \right| \leq \begin{bmatrix} \alpha \\ \gamma \end{bmatrix} \right\} \tag{29}$$

For brevity we shall refer to this set as the allowable set.

Expanding (25) and (27) in terms of α, γ we have

$$\Gamma_{11}\alpha + \Gamma_{12}\gamma \leq \alpha \quad \Gamma_{21}\alpha + \Gamma_{22}\gamma \leq \gamma \quad |\hat{K}_1\alpha| + |\hat{K}_2\gamma| \leq \bar{u} \tag{30a–c}$$

Condition (30a) indicates that Γ_{11} must be a contraction. As explained earlier this is consistent with the assumption that for $\underset{\rightarrow}{c} = 0$ (in the absence of constraints) the feedback controller K should cause the nonlinear feedback system to behave in an acceptable manner for all initial conditions x in the allowable set

$$\left| \hat{W} \begin{bmatrix} V_o x \\ 0 \end{bmatrix} \right| \leq \bar{z}$$

Acceptable behaviour of course implies that f itself must be contractive with respect to its first arguments. Both conditions (30a) and (30c) emphasise the trade-off that exists between the size of allowable sets and the amount of control authority that is made available through the perturbation c. Large allowable sets require large α and these, through (30a, c) imply small γ which in turn limits the control authority, and vice versa.

To conclude the section we combine the results of this section together with the arguments used in Section 7.3 to state without proof the following stability theorem.

Theorem 3: Providing that $\hat{W}$, W_o and Q are so chosen that Γ_{11} is a contraction in the sense that it satisfies (30a) for $\gamma = 0$ and that the initial condition is in the allowable set $P_x(\alpha, \gamma)$ for some $\underset{\rightarrow}{c}$, Algorithm 2 has guaranteed stability and will asymptotically converge to the control law $u = -Kx$.

The section below gives simple numerical examples which illustrate the application and features of Algorithms 1 and 2.

7.5 Numerical examples

7.5.1 Application of Algorithm 1

For simplicity we shall consider a polytopic uncertainty class defined by just two corner plants as described below:

$$a(z) = \mu a_1(z) + (1-\mu)a_2(z), \qquad b(z) = \mu b_1(z) + (1-\mu)b_2(z) \quad 0 \le \mu \le 1$$
$$a_1(z) = 1 - 1.74z^{-1} + 0.246z^{-2} + 0.246z^{-3}, \quad b_1(z) = 1.3404z^{-1} + 0.1149z^{-2} - 0.4528z^{-3}$$
$$a_2(z) = 1 - 1.86z^{-1} + 0.294z^{-2} + 0.294z^{-3}, \quad b_2(z) = 0.6596z^{-1} + 0.2851z^{-2} - 0.0272z^{-3}$$

The nominal model, obtained for $\mu = 0.5$, is open-loop unstable with poles at 1.5, 0.6, -0.3 and zeros at -0.6, 0.4. The constraints to be imposed on the inputs are

$$-1.5 \le u \le 1.5 \quad |\Delta u| \le 0.3$$

The nominal LQ controller, designed to minimise the cost of (2) for $\lambda = 1$, can be implemented by the feedback configuration of Figure 7.3 for

$$N_o(z) = 1.7723 - 1.6913z^{-1} + 0.0647z^{-2} + 0.2202z^{-3}$$
$$D_o(z) = 1 + 0.1057z^{-1} - 0.1098z^{-2}$$

For $Q = 0$ this controller does not stabilise the system over the whole uncertainty class and thus instability becomes a problem even in the absence of constraints. Robust stability can be achieved in the unconstrained case through the use of a nonzero Youla parameter. The benefit of a 4th-order polynomial

$$Q(z) = 0.0426 - 0.4605z^{-1} - 0.1517z^{-2} - 0.1952z^{-3} - 0.0932z^{-4}$$

designed to minimise the infinity norm of the sensitivity function is a norm reduction from 3 (for $Q = 0$) to 2. Although this choice of Q is not optimal with respect to the particular uncertainty class used here, it renders the feedback system of Figure 7.3 robustly stable and therefore can be used in the application of Algorithm 1. It is noted that, as explained earlier, in the unconstrained case, the insertion of Q does not affect the optimality of the LQ controller.

The simulation results for a unit setpoint change when the algorithm is applied to the nominal plant ($\mu = 0.5$) are shown in Figure 7.6 and are clearly satisfactory. The algorithm was effective in dealing with the infeasibility of the LQ controller as demonstrated by the nonzero value of the perturbation signal c (Figure 7.6*c*) which causes the control increments (shown in Figure 7.6*b*) to remain within their limits; naturally during transients these limits are active. As expected, Algorithm 1 achieves stability for the whole class while yielding satisfactory performance even for the corner models as demonstrated by the plots of Figure 7.7 for $\mu = 1$, and Figure 7.8 for $\mu = 0$. The output response of Figure 7.7*a* is satisfactory and not very dissimilar to that obtained for the nominal plant, though the input responses shown in Figure 7.7*b* are more active. The responses of Figure 7.8 are somewhat oscillatory, but that is hardly surprising when one considers the excessive amount of uncertainty allowed for. The key point is that the perturbation signal c once again coped well with infeasibility.

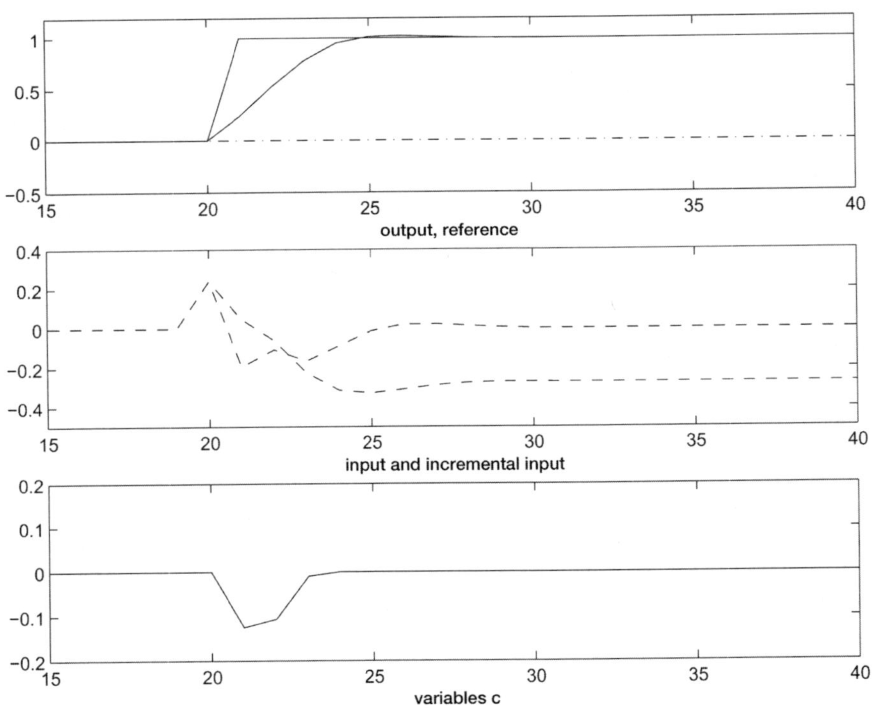

Figure 7.6 Simulation results for Algorithm 1 applied to nominal model ($\mu = 0.5$)

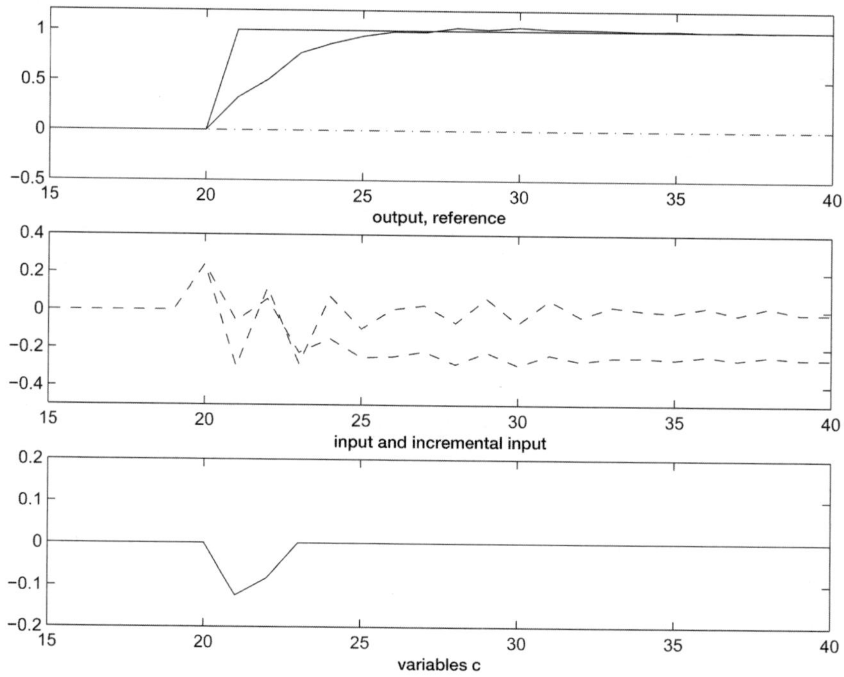

Figure 7.7 Simulation results for Algorithm 1 applied to the corner model for $\mu = 1$

7.5.2 Application of Algorithm 2

The example chosen here is based on the model of a coupled tank continuous time system which, after discretisation, reduces to:

$$\tilde{x}_{k+1} = \tilde{x}_k + A\Phi_o(\tilde{x}_k) + B\tilde{u}_k, \quad y_k = C\tilde{x}_k, \quad |u| \leq 0.5$$
$$A = \begin{bmatrix} -1.0361 & 0.3684 \\ 0.5181 & -0.3684 \end{bmatrix}, \quad B = \begin{bmatrix} 2.5029 \\ 0 \end{bmatrix}, C = [0 \quad 1]$$
$$\Phi_o(x) = \begin{bmatrix} \varphi([x]_1) \\ \varphi([x]_2) \end{bmatrix}, \quad \varphi([x]_i) = \text{sign}([x]_i)\sqrt{|[x]_i|}, \ i = 1, 2$$

where the elements of $\tilde{x}$ represent the difference in the two tank levels and the level in the second tank. The aim will be to steer the system to the equilibrium condition defined by

$$A\Phi_o(\tilde{x}^o) + B\tilde{u}^o = 0, \ y^o = 75 \quad \Leftrightarrow \quad \tilde{u}^o = 0.66, \tilde{x}^o = \begin{bmatrix} 67.5 \\ 75 \end{bmatrix}$$

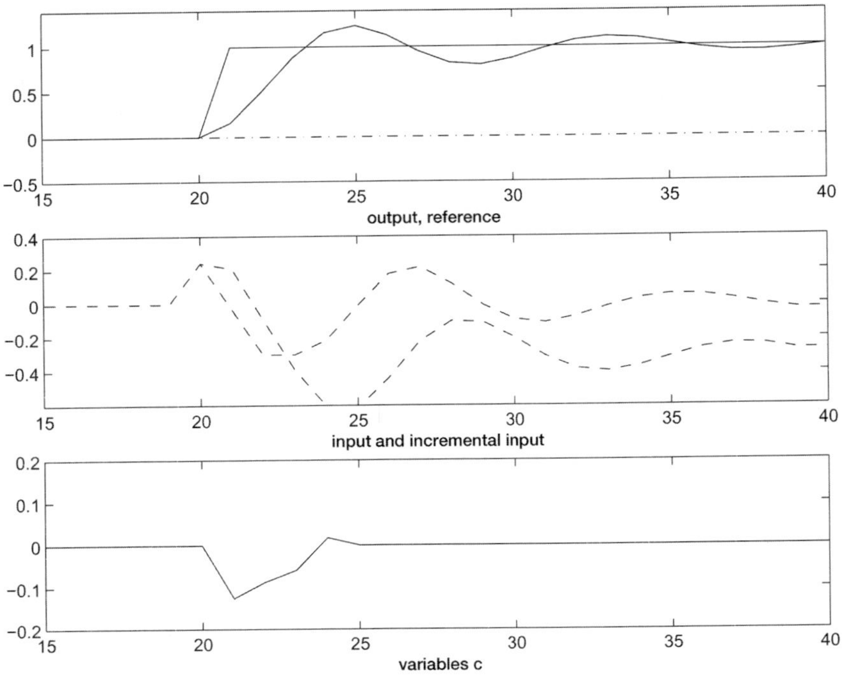

Figure 7.8 Simulation results for Algorithm 1 applied to the corner model for $\mu = 0$

which can be combined with the difference equation above to describe the error dynamics as:

$$x_{k+1} = x_k + A\Phi(x_k) + Bu_k, \quad e_k = Cx_k,$$
$$x_k = \tilde{x}_k - \tilde{x}^o, \quad e_k = y_k - y^o, \quad u_k = \tilde{u}_k - \tilde{u}^o$$
$$\Phi(x) = \begin{bmatrix} \varphi_1([x]_1) \\ \varphi_2([x]_2) \end{bmatrix}, \quad \varphi_i([x]_i) = \varphi_o([x]_i + [x^o]_i) - \varphi_o([x^o]_i), \quad i = 1, 2$$

The nominal state feedback controller K (see equation 16) is chosen to be the LQR optimal controller minimising the cost $\|x\|_2^2 + \|u\|_2^2$ evaluated for the above model after linearisation about the origin and is given by

$$K = [0.2985 \quad 0.2155]$$

Introduction of the perturbed optimal controller law of (16) into the nonlinear error model above, and taking $W_o = I$, gives the autonomous system description of (20) as

$$z_{k+1} = \begin{bmatrix} f(q_k, E\underset{\rightarrow}{c_k}) \\ T\underset{\rightarrow}{c_k} \end{bmatrix}, \quad f(q, E\underset{\rightarrow}{c}) = (I + AD - BK)q + A[\Phi(q) - Dq] + BE\underset{\rightarrow}{c}$$

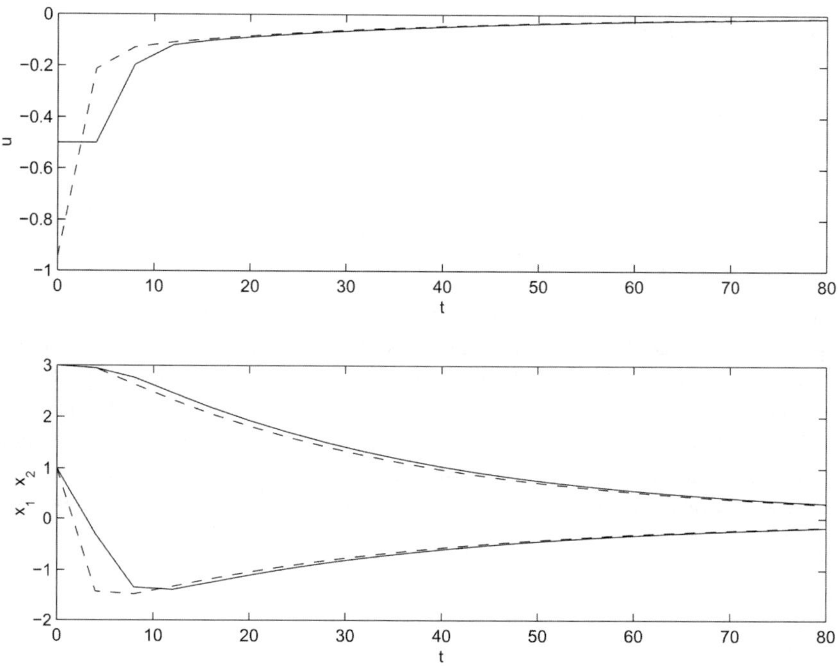

Figure 7.9 Closed-loop responses of Algorithm 2 (solid line) and unconstrained LQ feedback (dashed line) for initial condition $x = [1\ 3]$

Then the polyhedral set of (21), (24) can be determined on the basis of the gain bounds:

$$|\Phi(q) - Dq| \leq \begin{bmatrix} 0.1212 & 0 \\ 0 & 0.0862 \end{bmatrix} |x| \quad \forall q = x \in \Pi; \quad \Pi = \left\{ x : |x| \leq \begin{bmatrix} 5.8354 \\ 11.5397 \end{bmatrix} \right\}$$

for

$$\left| \hat{W} \begin{bmatrix} f(q, Ec) \\ T\underset{\rightarrow}{c} \end{bmatrix} \right| = \left| \hat{W} \begin{bmatrix} I + AD - BK & BE \\ 0 & T \end{bmatrix} \hat{V} z + \hat{W} \begin{bmatrix} A[\Phi(q) - Dq] \\ 0 \end{bmatrix} \right|$$

$$\leq \left\{ \left| \hat{W} \begin{bmatrix} I + AD - BK & BE \\ 0 & T \end{bmatrix} \hat{V} \right| + \left| \hat{W} \begin{bmatrix} A \\ 0 \end{bmatrix} \right| \times \begin{bmatrix} 0.1212 & 0 \\ 0 & 0.0862 \end{bmatrix} |\hat{V}[I \quad 0]| \right\} |z| = \Gamma |z|$$

Using this relationship the volume of the polyhedral set $P_x(\alpha, \gamma)$, for one degree of freedom, can be maximised numerically subject to the invariance/feasibility constraints (26), (28) to give:

$$\hat{W} = \begin{bmatrix} 1 & 0.7221 & -3.3503 \\ -0.9666 & 1 & -4.6396 \\ 0.0236 & 0.0815 & 1 \end{bmatrix}, \quad \bar{z} = \begin{bmatrix} 1.6751 \\ 11.4021 \\ 0.4560 \end{bmatrix},$$

$$\Gamma = \begin{bmatrix} 0.0505 & 0.0382 & 2.5326 \\ 0.8573 & 0.7207 & 3.0694 \\ 0.0295 & 0.0250 & 0.2656 \end{bmatrix}$$

With these parameters the application of Algorithm 2, as expected, produced (Figure 7.9*a*,*b*) satisfactory performance (solid lines) especially when compared with the unconstrained optimal performance (dashed lines). Figure 7.10, on the other hand, shows (in solid lines) the set $P_x(\alpha, \gamma)$ for one degree of freedom. This set clearly covers a greater part of Π (indicated by the rectangular box drawn in dotted lines) than the maximal admissible set for zero degrees of freedom (drawn in dash-dotted lines) and a comparable part as the maximal admissible set for one

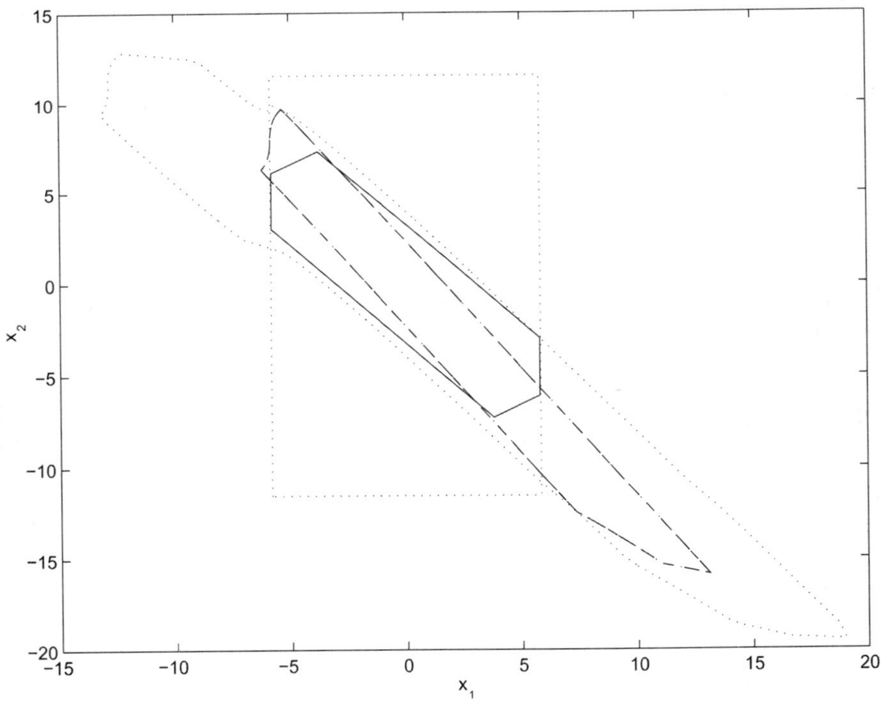

Figure 7.10 *Initial conditions sets for example in Section 7.5.2 (rectangle in dotted lines – set of initial conditions; solid lines – polytope $P_x(\alpha, \gamma)$; dash-dotted lines – maximal admissible set for zero degrees of freedom; dotted lines – maximal admissible set for one degree of freedom)*

degree of freedom (shown in dotted lines). It is noteworthy that this favourable comparison with the maximal admissible sets is achieved despite the approximation involved in the use of the linear upper bounds on $|\Phi(q) - Dq|$.

7.6 Acknowledgment

The authors wish to thank the EPSRC for financial support.

7.7 References

1 CLARKE, D.W., MOHTADI, C., and TUFFS, P.S.: 'Generalised predictive control: Parts 1 and 2', *Automatica*, 1987, **23** (2), pp. 137–60

2 KWON, W.H., and PEARSON, A.E.: 'On feedback stabilisation of time-varying discrete linear systems', *IEEE Trans. Autom. Control.*, 1978, **AC-23** (3), pp. 479–81

3 KEERTHI, S.S., and GILBERT, E.G.: 'Optimal infinite-horizon feedback laws for a class of constrained discrete-time systems. Stability and moving horizon approximations', *J. Optimiz. Theory Appl.*, 1988, **57** (2), pp. 265–93

4 CLARKE, D.W., and SCATTOLINI, R.: 'Constrained receding horizon predictive control', *IEE Proc. D*, 1991, **138** (4), pp. 347–54

5 MOSCA, E., and ZHANG, J.: 'Stable redesign of predictive control', *Automatica*, 1992, **28** (6), pp. 1229–33

6 KOUVARITAKIS, B., ROSSITER, J.A., and CHANG, A.O.T.: 'Stable generalised predictive control: an algorithm with guaranteed stability', *IEE Proc. D*, 1992, **139** (4), pp. 349–63

7 RAWLINGS, J.B., and MUSKE, K.R.: 'The stability of constrained receding horizon control', *IEEE Trans. Autom. Control*, 1993, **AC-38** (10), pp. 1512–6

8 ROSSITER, J.A., GOSSNER, J.R., and KOUVARITAKIS, B.: 'Infinite horizon stable predictive control', *IEEE Trans. Autom. Control*, 1996, **AC-41** (10), pp. 1522–7

9 ROSSITER, J.A., KOUVARITAKIS, B., and RICE, M.J.: 'A numerically robust state-space approach to stable predictive control strategies', *Automatica*, 1998, **34** (1), pp. 65–73

10 KOUVARITAKIS, B., ROSSITER, J.A., and CANNON, M.: 'Linear quadratic feasible predictive control', *Automatica*, 1998, **34** (12), pp. 1583–92

11 SZNAIER, M., and DAMBORG, M.J.: 'Suboptimal control of linear systems with state and control inequality constraints'. Proc. IEEE Conference on Decision and control, 1987, pp. 761–2

12 SCOKAERT, P.O.M., and RAWLINGS, J.B.: 'Infinite horizon linear quadratic control with constraints'. Proceedings of IFAC World Congress, vol. M, San Francisco, 1996, pp. 109–14

13 CHMIELEWSKI, D., and MANOUSIOUTHAKIS, V.: 'On constrained infinite-time linear quadratic control', *Syst. Control Lett.*, 1996, **29**, pp. 121–9

14 MICHALSKA, H., and MAYNE, D.Q.: 'Robust receding horizon control of constrained non-linear systems', *IEEE Trans. Autom. Control*, 1993, **AC-38** (11), pp. 1623–33
15 CHEN, H., and ALLGÖWER, F.: 'A quasi-infinite horizon non-linear model predictive control scheme with guaranteed stability', *Automatica*, 1998, **34** (10), pp. 1205–17
16 ROSSITER, J.A., and KOUVARITAKIS, B.: 'Numerical robustness and efficiency of generalised predictive control algorithms with guaranteed stability', *IEE Proc. D*, 1994, **141** (3), pp. 154–62
17 ROSSITER, J.A., and KOUVARITAKIS, B.: 'Youla parameter and robust predictive control with constraint handling'. International symposium on *Non-linear model predictive control*, Ascona, 1998
18 [1]KOUVARITAKIS, B., ROSSITER, J.A., and SCHUURMANS, J.: 'Efficient robust predictive control', *IEEE Trans. Autom. Control.*, 2000, **45** (8), pp. 1545–9
19 [2]LEE, Y.I., and KOUVARITAKIS, B.: 'Stabilisable regions of receding horizon predictive control with input constraints', *Syst. Control Lett.*, 1999, **38** (1), pp. 13–20
20 [3]LEE Y.I., and KOUVARITAKIS, B.: 'Linear matrix inequalities and polyhedral invariant sets in constrained robust predictive control', *Int. J. Non-linear Robust Control*, 2000, **10** (13), pp. 1079–90
21 GILBERT, E.G., KOLMANOVSKY, I., and TAN, K.T.: 'Non-linear control of discrete-time linear systems with state and control constraints: a reference governor with global convergence properties'. Proceedings of CDC, Florida, 1994, pp. 144–9
22 BEMPORAD, A., and MOSCA, E.: 'Constraint fulfilment in control systems via predictive reference management'. Proceedings of CDC, Florida, 1994, pp. 3017–22
23 GOSSNER, J.R., KOUVARITAKIS, B., and ROSSITER, J.A.: 'Stable generalised predictive control wth constraints and bounded disturbances', *Automatica*, 1997, **33** (4), pp. 551–68
24 SCOKAERT, P.O.M., and CLARKE, D.W.: 'Stability and feasibility in constrained predictive control', in 'Advances in model-based predictive control' (Oxford University Press, 1994)
25 KOTHARE, M.V., BALAKRISHNAN, V., and MORARI, M.: 'Robust constrained model predictive control using linear matrix inequalities', *Automatica*, 1996, **32** (10), pp. 1361–79

Chapter 8

Computationally efficient nonlinear model predictive control algorithm for control of constrained nonlinear systems

Alex Zheng and Wei-hua Zhang

Abstract

A nonlinear model predictive control algorithm is proposed for control of constrained nonlinear systems. The basic idea is to calculate exactly the first control move, which is implemented, and approximate all other control moves, which are not implemented. Regardless of the control horizon, the number of decision variables for the on-line optimisation problem equals the number of manipulated variables, resulting in significant savings in on-line computational time. The feasibility for a practical implementation of the proposed algorithm is demonstrated on two examples including the Tennessee-Eastman Challenge process involving seven inputs and three outputs.

8.1 Introduction

Following successful industrial application of linear model based control algorithms, various nonlinear model predictive control (NMPC) algorithms have been proposed for control of nonlinear systems with constraints (see, for example, References 1–7). Several excellent reviews have been published (see, for example, References 8 and 9). In general, NMPC solves an on-line optimisation problem in real time at each sampling time. In terms of how the on-line optimisation problem is solved, the algorithms can be divided into two main groups – one attempts to solve the optimisation problem exactly (exact optimisation approach) and the other tries to solve the problem via (usually linear) approximations (approximation approach).

The main advantages for the exact optimisation approach are that performance is 'optimal' and that strong and nonconservative nominal stability results have been proven (see, for example, References 7, 10 and 11) and have been nicely reviewed by De Nicolao *et al.* [12]. Since the on-line optimisation problem is generally nonconvex, the on-line computational demand is high for any reasonably nontrivial systems. Thus, practical implementation of the exact optimisation approach is difficult, if not impossible. One may argue that the problem with high on-line computational demand would vanish as faster and cheaper computers and more powerful software become available. This may be true for small problems but unrealistic for large problems due to exponential growth in computational time with respect to the number of decision variables (which equals the number of inputs times the control horizon).

Due to the prohibitively high on-line computational demand associated with the exact optimisation approach, García [1] proposed to linearise the nonlinear model successively. Gattu and Zafiriou [13, 14], and Lee and Ricker [5] incorporated the state estimation techniques into this approach. By introducing a stability constraint and assuming the feasibility of the optimisation problem, De Oliveira and Morari [15] have proposed an NMPC algorithm with guaranteed robust stability. Another approach [16, 17] is to linearise (either input/output or feedback) the nonlinear system. For a certain class of nonlinear systems, such transformation yields a linear dynamic system but with state-dependent (and, in general, nonlinear) constraints which need to be approximated in order to generate a computationally simpler optimisation problem (otherwise, these methods would fall into the exact optimisation approach and suffer the same disadvantage computationally). An attractive feature of this approach is that only a quadratic program needs to be solved on-line in real time. However, the design may be overly conservative in some cases: linear approximation is only valid when both state and input do not deviate too much from where they are linearised. This implies that the control actions have to be close to their linearised values in order to preserve stability. Thus we may be sacrificing performance for computational simplicity.

In this chapter we propose an NMPC algorithm which combines the best features of the exact optimisation approach (thus performance) and the approximation approach (thus low on-line computational demand). The key idea behind the algorithm is to calculate exactly the first control move, which is actually implemented, and to approximate the rest of the control moves, which are not implemented. Thus the number of decision variables in the on-line optimisation problem, regardless of the control horizon, equals the number of inputs, instead of the number of inputs times the control horizon for a conventional NMPC algorithm. This results in significant savings in on-line computational demand since the control horizon needs to be chosen reasonably large to ensure adequate performance and in some cases stability, and since the optimisation problem is generally nonconvex.

8.2 Preliminaries

We consider a continuous-time nonlinear system described as follows (the discrete-times system description was treated in Reference 18):

$$\begin{aligned} \dot{x}(t) &= f(x(t), u(t)) \\ y(t) &= g(x(t)) \end{aligned} \tag{1}$$

where $x \in X \subset \Re^{n_x}$ is the state, $u \in \mathrm{U} \subset \Re^{n_u}$ is the input, and $y \in \mathrm{Y} \subset \Re^{n_y}$ is the output. The sets U and Y are given as follows:

$$\begin{aligned} \mathrm{U} &\triangleq \{u : 0 > u^{\min} \leq u \leq u^{\max} > 0\} \\ \mathrm{Y} &\triangleq \{y : 0 > y^{\min} \leq y \leq y^{\max} > 0\} \end{aligned} \tag{2}$$

The objective function at sampling time k is defined as

$$\begin{aligned} \Phi_k = &\int_0^{H_p} (r(k+t) - y(k+t|k))^T \Gamma_y(y(k+t|k))(r(k+t) - y(k+t|k))dt \\ &+ \sum_{i=0}^{H_c-1} \Delta u(k+i|k)^T \Gamma_u \Delta u(k+i|k) \end{aligned} \tag{3}$$

where H_p is the output horizon, H_c is the control (or input) horizon, k is the kth sampling time (the sampling period is assumed to be one for simplicity), r is the desired output trajectory, $\Delta u(k)\ u(k) - u(k-1)$, $\bullet(k+i|k)$ denotes the variable at time $k+i$ computed at time k, and $\Gamma_y(y) = \Gamma_y^T(y) \geq \Gamma_y(0) \geq 0$ and $\Gamma_u = \Gamma_u^T > 0$ are output and input weights, respectively. The output weight $\Gamma_y(y(k+t|k))$ is assumed to be diagonal and depends on the predicted output $y(k+t|k)$. The reason behind it as well as its functional form will be discussed shortly. Notice that the inputs are discrete (i.e. $u(t)$ is constant for $t \in [k, k+1)$) while the outputs are continuous. Δu is penalised instead of u to generate integral control. Now we define Controller NMPC 1, which directly extends the DMC algorithm for linear systems [19] as follows.

Definition 1 (*Controller NMPC 1*): *At each sampling time k, $\Delta u(k)$ equals the first control move $\Delta u(k|k)$ of the sequence $\{\Delta u(k|k)), \ldots, \Delta u(k+H_c-1|k)\}$:*

$$\min_{\Delta u(k|k),\ldots,\Delta u(k+H_c-1|k)} \Phi_k \text{ subject to } \begin{cases} \dot{x}(k+t|k) = f(x(k+t|k), u(k+t|k)), t \geq 0 \\ y(k+t|k) = g(x(k+t|k)), t \geq 0 \\ u(k+i|k) \in U, i = 0, \ldots, H_c - 1 \\ \Delta u(k+1|k) = 0, i \geq H_c \end{cases} \tag{4}$$

where the objective function Φ_k is defined by (3).

Notice that the output constraints are not included in the optimisation problem. We have chosen to soften the output constraints through the output-dependent weight

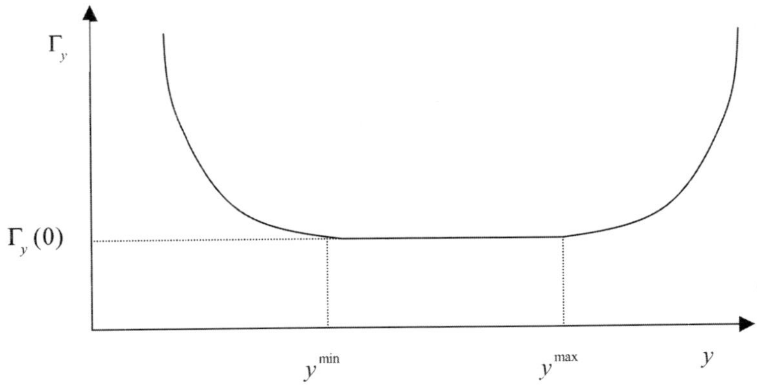

Figure 8.1 Output weight ($\Gamma_y(y)$) versus ouput (y)

$\Gamma_y(y)$, which has the following form (Figure 8.1):

$$\Gamma_y(y) = \begin{bmatrix} \Gamma_{y1}(y_1) & 0 & \dots & 0 \\ 0 & \Gamma_{y2}(y_2) & \dots & 0 \\ & \vdots & \ddots & \\ 0 & \dots & 0 & \Gamma_{yny}(y_{ny}) \end{bmatrix},$$

$$\Gamma_{yi}(y_i) = \begin{cases} \Gamma_{yi}(0)\left[1 + \varepsilon_i\left(y_i - y_i^{\min}\right)^2\right] & \text{if } y_i \le y_i^{\min} \\ \Gamma_{yi}(0) & \text{if } y_i^{\min} \le y_i \le y_i^{\max} \\ \Gamma_{yi}(0)\left[1 + \varepsilon_i\left(y_i - y_i^{\max}\right)^2\right] & \text{if } y_i \ge y_i^{\max} \end{cases} \qquad (5)$$

Here $\varepsilon_i \ge 0$ can be chosen to achieve the desirable degree of softening: $\varepsilon_i = 0$ indicates no constraints while $\varepsilon_i = \infty$ indicates hard constraints. It can be easily shown that this is equivalent to the output softening discussed in References 20 and 21. Clearly other functional forms can also be used.

Unlike dynamic matrix control, where the on-line optimisation problem is a quadratic program, the optimisation problem (5) is in general nonconvex. In the worst case, the on-line computational demand increases exponentially with the number of decision variables (i.e. $n_u H_c$). For reasons of stability and performance, H_c should be chosen reasonably large: it is well known that, for a linear system with n unstable poles (including those on the $j\omega$ axis for a continuous-time system or on the unit circle for a discrete-time system), closed-loop stability is guaranteed only if $H_c \ge n$ [22, 23]. Also, as we demonstrate in Example 2 of this chapter, choosing small H_c may result in worse performance than that achievable by a linear controller. Therefore, on-line implementation of Controller NMPC 1 with a reasonable value of H_c is difficult for any reasonably large system. The question is:

Can we significantly reduce the on-line computational time for any H_c with minimal loss in performance?

8.3 Computationally efficient algorithm

Since all future control moves $\Delta u(k+1|k), \ldots, \Delta u(k+H_c-1|k)$ are not implemented, the following question naturally arises: If we do not have sufficient time to solve for all of them on-line, how should we approximate them 'optimally'? Suppose we only have enough time to compute $\Delta u(k|k)$ on line. Let us first focus on approximating $\Delta u(k+1|k)$ (or equivalently $u(k+1|k)$). The optimal value of $u(k+1|k)$ is determined by solving the optimisation problem (5). Suppose $u(k+1|k) = h(x(k+1|k), e(k))$, where $e(k+1|k) = r(k+1) - y(k+1|k)$. The controller $h(x, e)$ is generally nonlinear and is impossible to compute analytically, except for some trivial nonlinear systems. However, the controller would be linear and can be computed analytically off-line if the system were linear and there were no constraints [24]. Thus, an optimal linear approximation of $u(k+1|k)$ can be obtained by linearising the nonlinear system at $x(k+1|k)$ and $u(k+1|k)$ and by assuming no constraints (this implies that Γ_y is constant), i.e. $u(k+1|k) \approx C(x(k+1|k), e(k+1|k))$, where C is the linear controller corresponding to Controller NMPC 1 with a linear model (linearised at $\{x(k+1|k), u(k|k)\}$) and without constraints. In other words, the linear approximation of $\Delta u(k+1|k)$ equals what would be implemented if the system were linear and there were no constraints. Following similar arguments, linear controllers which optimally approximate future control moves would correspond to Controller NMPC 1 with linear models (linearised at appropriate points) and without constraints. Thus, such approximation would allow us to compute $\Delta u(k|k)$ as follows:

Step 0: Guess a value for $\Delta u(k|k)$ and set $i = 1$.
Step 1: Compute $x(k+i|k)$ and linearise the nonlinear system at $x(k+i|k)$ and $u(k+i-1|k)$.
Step 3: Set $i = i + 1$ and repeat Steps 1–2 if $i \leq H_p$.
Step 4: Update $\Delta u(k|k)$, set $i = 1$ and repeat Steps 1–3 until an optimal solution has been found.

While such linearisation and analytical determination of the linear controllers are straighforward, they require linearising the system on-line during the optimisation over $\Delta u(k|k)$, which can be computationally expensive, since $x(k+i|k)$ depend on $u(k+j|k), j < i$. A more computationally efficient approach is to divide the region of all possible operations into N subregions, linearise the nonlinear system at one point in each of these N regions, and determine N linear controllers for the N linear models (one for each region) off-line. These N linear controllers can be stored for on-line use. For simplicity, we have chosen N to be one for both examples discussed in Section 4.

Controller NMPC 2, defined below, assumes that future control moves can be approximated by a controller C (i.e. $u(k+i|k) = \text{sat}\,(C(x(k+i|k), e(k+i|k)), i = 1, \ldots, H_p - 1)$.

Definition 2 (*Controller NMPC 2*): *At each sampling time k, $\Delta u(k)$ is the solution to the following optimisation problem:*

$$\min_{\Delta u(k|k)} \Phi_k \quad \text{subject to} \begin{cases} \dot{x}(k+t|k) = f(x(k+t|k), u(k+t|k)), t \geq 0 \\ y(k+t|k) = g(x(k+t|k), u(k+t|k)), t \geq 0 \\ u(k|k) \in U \\ u(k+1|k) = \text{sat}(C(x(k+i|k), e(k+i|k))), \\ \qquad i = 1, \ldots, H_p - 1 \end{cases} \tag{6}$$

where sat denotes the saturation function, i.e.

$$\text{sat}(u) = \begin{Bmatrix} \text{sat}(u_1) \\ \vdots \\ \text{sat}(u_{n_u}) \end{Bmatrix}, \text{sat}(u_i) = \begin{cases} u_i^{\max} & u_i > u_i^{\max} \\ u_i & u_i^{\min} \leq u_i \leq u_i^{\max} \\ u_i^{\min} & u_i < u_i^{\min} \end{cases}$$

The on-line optimisation problem for Controller NMPC 2 has only n_u decision variables (versus $n_u\, H_c$ for Controller NMPC 1), *regardless* of the control horizon H_c. This results in significant savings in on-line computational time as the on-line computational time grows exponentially with the number of decision variables in the worst case. Another difference is that $H_c = H_p$ for Controller NMPC 2. This is because the control horizon has no effect on the on-line computational demand for Controller NMPC 2. Also the control parameterisation strategy differs: Controller NMPC 1 parameterises the future control moves (i.e. $\Delta u(k+i|k), i \geq 1$) as a function of time while Controller NMPC 2 parameterises them as a function of state. In the nominal case (i.e. perfect model) there is no difference between the two parameterisations. In the presence of modelling error, however, there is a significant difference between the two and parameterizing the future control moves as a function of state is preferred [25–27]. To see this difference, let us consider a simple example.

Example 1: Consider the following system:

$$y(k) = au(k-1), \quad 0.8 \leq a \leq 1, \quad y(0) = 0, u(t) = 0 \forall t < 0$$

The setpoint is one and the objective to be minimised at each sampling time is

$$\Phi_k = \max_a \sum_{i=1}^{\infty} (1 - y(k+i))^2$$

At sampling time k, an open-loop optimal control strategy would determine a fixed sequence of controls such that the objective function is minimised. Regardless of what the control sequence is, Φ_k is not bounded. However, if we parameterise the future control moves as a function of future state (e.g. $u(k+i) = u(k+i-1) + b(1 - y(k+i)), i \geq 1$ with b appropriately chosen), Φ_k would be bounded if $u(k)$ is chosen properly.

Remark 1: *It should be emphasised that while we have restricted ourselves in constructing linear discrete-time controllers which provide optimal linear approximations of $u(k+i|k), i = 1, \ldots, H_p - 1$, in general, C can be nonlinear and/or continuous.*

Remark 2: *For simplicity, Controller NMPC 2 assumes that the future inputs are clipped to satisfy the constraints via the saturation function, which may result in poor performance, especially for ill-conditioned multivariable systems. Other anti-windup structures should be used instead (for example, References 28 and 29).*

Remark 3: *Controller NMPC 2 assumes that the allowable time is only sufficient to compute the first control move. Extension to computing multiple control moves, when possible, is straightforward.*

8.4 Examples

The proposed algorithm is illustrated on a distillation column dual composition control problem (LV-configuration) using a rigorous tray-by-tray model with 82 states [30], and the TE Challenge problem [31] involving 7 inputs and 3 outputs. We emphasise that the sole purposes of these examples are to compare Controllers NMPC 1 and 2 and to demonstrate that it is feasible, as far as the on-line computational effort is concerned, to implement Controller NMPC 2 on a large nonlinear system. It is not our purpose to show that nonlinear controllers are needed to control these systems. For simplicity, we have used one linear controller for C in both examples.

8.4.1 Distillation dual composition control

The tray-by-tray model developed by Skogestad [30] is used as both the nominal model and actual plant (i.e. no model uncertainty). The sampling period is 5 min. Theoretically, an infinite output horizon needs to be used to guarantee nominal stability. Since an infinite output horizon would require an infinite amount of time to compute the objective function value and is impossible to implement, a large output horizon ($H_p = 400$, i.e. 2000 min) is used instead (the open-loop settling time is about 800 min). Figure 8.2 shows the on-line computational time per sampling time for determining one local solution versus H_c for the two algorithms for $H_p = 400$ based on actual simulations. (It should be emphasised that the absolute values in the figure are not crucial as they depend on initial guess,

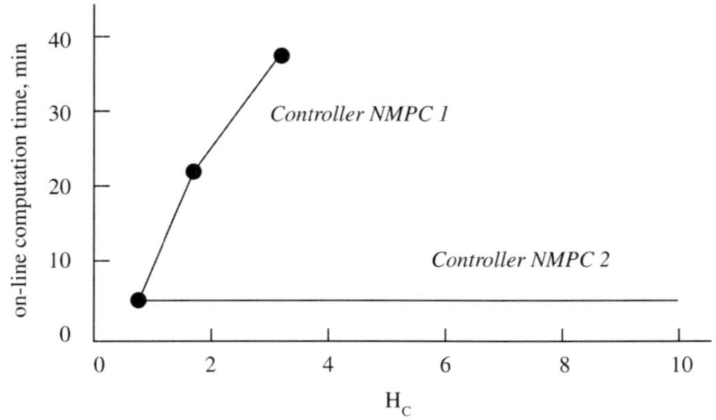

Figure 8.2 Comparison of on-line computational time per sampling time for determining one local solution for Controllers NMPC 1 and 2 ($H_p = 400$). The on-line optimisation problem is solved using a conjugate gradient method in MATLAB/ SIMULINK on a Pentium 266 mHz PC. The integration routine gear is used

software, hardware, *etc.* What is important is the trend.) Regardless of H_c, the computational time for Controller NMPC 2 is about 3 min, which is larger than the computational time for Controller NMPC 1 with $H_c = 1$ (about 2.5 min) because the inclusion of the controller C increases the time for evaluating the objective function which involves simulation. In general, since the nonlinear system is much more complex than the controller C, simulation time for the nonlinear system with or without C should be very similar. Therefore, including the controller C should have a very small impact on on-line computational time. The discrete controller C is determined analytically off-line based on the following linear model obtained at the nominal operating condition (i.e. distillation composition: 0.99; bottoms composition: 0.01):

$$G(s) = \frac{1}{190s + 1}\begin{bmatrix} 0.878 & -0.864 \\ 1.082 & -1.906 \end{bmatrix}$$

and the following tuning parameters:

$$\Gamma_y = I, \Gamma_u = 0.1I, H_c = 400, H_p = 400.$$

However, to speed up the simulations, a continuous controller is used for C.

It is clearly impossible to implement Controller NMPC 1 for the above tuning parameters. One may be tempted to use a smaller control horizon that is feasible (e.g., $H_c = 1$). This approach may not be desirable, as discussed below. Figure 8.3 compares the performance for Controller NMPC 2, Controller NMPC 1 with $H_c = 1$, and the linear controller C for a setpoint change in distillate composition from 0.99 to 0.999. The reason for such a big setpoint change is to illustrate the

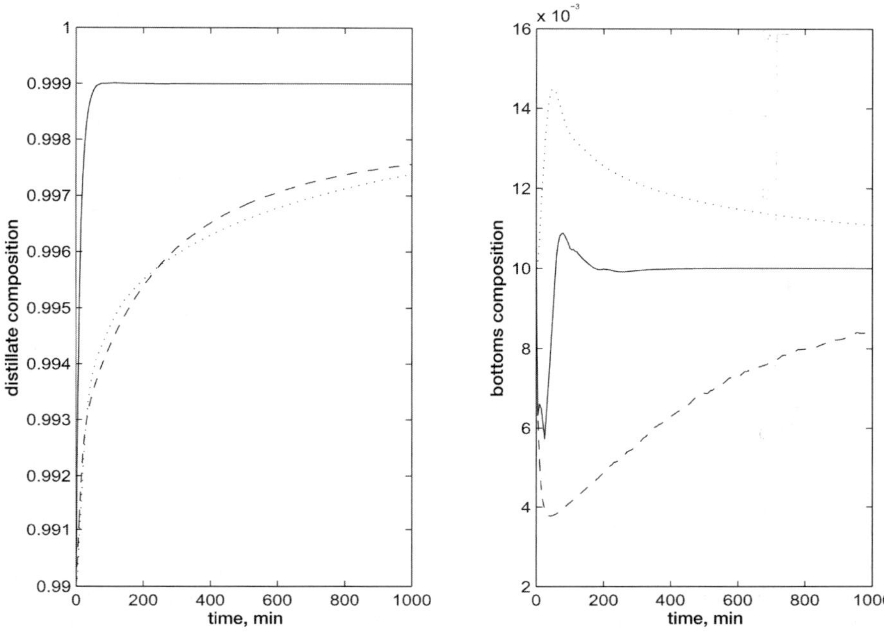

Figure 8.3 Performance for Controller NMPC 2 (solid), Controller NMPC 1 with $H_c = 1$ (dashed), and the linear controller C (dotted)

effect of nonlinearity. While the on-line computational times for Controller NMPC 2 and Controller NMPC 1 with $H_c = 1$ are about the same, the performance for Controller NMPC 2 is significantly better. Furthermore, Controller NMPC 1 with $H_c = 1$ does not seem to perform as well as the linear controller C, which requires almost no on-line computation; thus, simply reducing H_c to make the implementation feasible is in general not a good idea.

8.4.2 Tennessee-Eastman problem

The TE Challenge problem was proposed by Downs and Vogel [31] for testing alternative control and optimisation strategies for continuous chemical processes (Figure 8.4). A simplified model, which has 26 states and includes two PI controllers, developed by Ricker [32] is used as both the nominal model and the actual plant (i.e. no model uncertainty). (The two PI controllers control reactor and separator temperatures by manipulating reactor and condenser coolant valves, respectively.) The sole purpose of this example is to illustrate that it is feasible, as far as the on-line computational effort is concerned, to implement Controller NMPC 2 on nonlinear systems as large as the TE Challenge process.

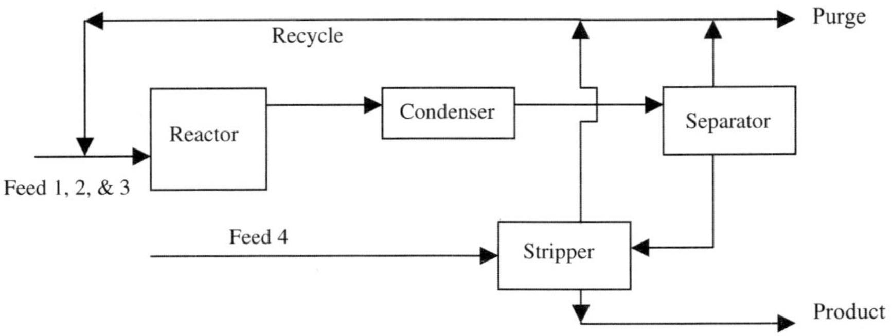

Figure 8.4 Simplified block diagram of the TE problem

Ten inputs are selected as manipulated variables and six outputs as controlled variables. They are listed in Table 8.1. Three level controllers, with parameter values from Reference 32, are implemented to reduce the size of the on-line optimisation problem for Controller NMPC 1 with large values of H_c. Thus, NMPC algorithms handle a subsystem with seven inputs and three outputs, which are italicised in Table 8.1. A sampling period of 5 min is used as all the controlled variables are available continuously with Ricker's model, which is not the case for the original model.

The optimisation problem is found to depend on the initial guess, implying a nonconvex optimisation problem. Solving for the global solution by searching over the allowable space was not 'feasible' for any value of H_c ('feasible' means that a simulation can be finished within a few days on a Pentium 266 MHz PC). We decide to approximate the global solution by the best solution among local solutions obtained from different initial guesses. The number of initial guesses needed to ensure that on average the best local solution is within 5 per cent of the global solution is determined by trial-and-error. It is found to depend on H_c. This method is only feasible for $H_c \leq 2$, implying that Controller NMPC 1 can only be

Table 8.1 Summary of manipulated and controlled variables

Manipulated variables	*Controlled variables*
Feed 1 (pure A), u_1	Reactor pressure, y_1
Feed 2 (pure D), u_2	Reactor liquid level
Feed 3 (pure E), u_3	Separator liquid level
Feed 4 (A), u_4	Stripper reboiler level
Recycle flow, u_5	H in product, y_2
Purge, u_6	Production rate, y_3
Separator underflow	
Stripper underflow	

implemented for $H_c = 1$ and 2, but Controller NMPC 2 can be implemented for all values of H_c. For $H_c = 1$, five initial guesses seem sufficient.

The following tuning parameters are used for both controllers:

$$H_c = 600 \text{ sampling periods}$$
$$\Gamma_u = 40\,\mathrm{diag}([90 \quad 15 \quad 12 \quad 4 \quad 1 \quad 66 \quad 100])$$
$$\Gamma_y = \mathrm{diag}([1 \quad 400 \quad 20])$$

The control horizon H_c equals 2 and 600 sampling periods for Controller NMPC 1 and Controller NMPC 2, respectively. Ideally we want to use $H_c = H_p$ for computing the linear controller C; however, this value was too big for the MPC Toolbox [33] to handle. We decide to use $H_c = 2$ to illustrate the performance loss (if any) of Controller NMPC 2 as compared to Controller NMPC 1 for the same value of H_c. After the controller C is determined via the MPC Toolbox, it is approximated by a continuous controller to speed up simulations.

Figure 8.5 shows performance for the two controllers for a setpoint change in per cent H in the product stream from 43.83 to 53.83 while maintaining the production rate at 211.3 kmoles/h and reactor pressure at 2700 kPa.

As one can see, it is difficult to judge which controller performs better, although the on-line computational time for Controller NMPC 2 is an order of magnitude less than that for Controller NMPC 1. Figure 8.6 compares the performance for the two controllers for a 10 per cent decrease in the composition of component B while

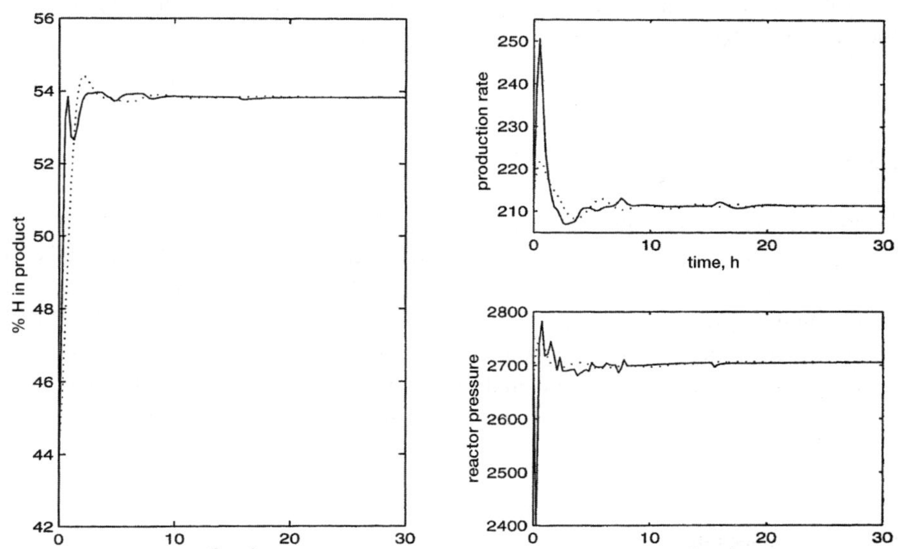

Figure 8.5 Servo responses for Controller NMPC 1 (solid) and Controller NMPC 2 (dotted)

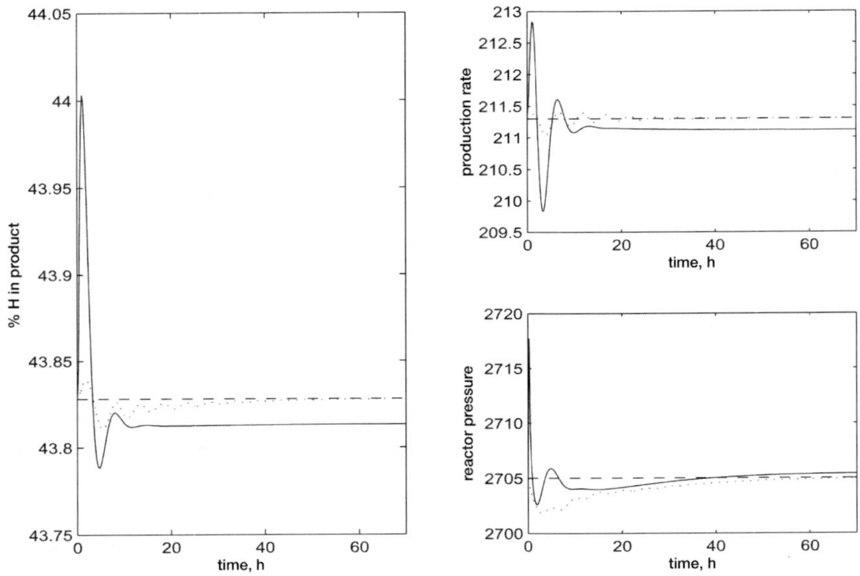

Figure 8.6 Regulatory responses for Controller NMPC 1 (solid) and Controller NMPC 2 (dotted)

maintaining the feed ratio of component A to C constant for Feed 4. Controller NMPC 2 performs better than Controller NMPC 1. In fact, the closed-loop system with Controller NMPC 1 has a steady-state offset, albeit a small one. It is entirely possible that the steady-state offset is due to local minima. However, it is possible for Controller NMPC 1 to have a steady-state offset [34].

8.5 Conclusions

In this chapter we have proposed a novel NMPC algorithm (Controller NMPC 2) for control of large nonlinear constrained systems. The number of decision variables for the on-line optimisation problem equals the number of inputs, instead of the number of inputs times the control horizon for a conventional NMPC algorithm, resulting in significant reduction in on-line computational time. The feasibility for a practical implementation of the algorithm has been demonstrated on the TE challenge process involving seven inputs and three outputs.

8.6 Acknowledgment

The authors gratefully acknowledge the financial support of NSF CAREER program.

8.7 References

1 GARCÍA, C.E.: 'Quadratic dynamic matrix control of nonlinear processes. An application to a batch reactor process'. AIChE Annual Meeting, San Francisco, CA, 1984

2 JANG, S., JOSEPH, B., and MUKAI, H.: 'On-line optimisation of constrained multivariable chemical processes', *AIChE Journal*, 1987, **33**, pp. 26–42

3 MORSHEDI, A.M.: 'Universal dynamic matrix control'. Proceedings of 3rd international conference on *Chemical process control* (CPC III), New York, 1986

4 EATON, J.W., and RAWLINGS, J.B.: 'Feedback control of nonlinear processes using on-line optimisation techniques', *Comput. Chem. Eng.*, 1990, **14**, pp. 469–79

5 LEE, J.H., and RICKER, N.L.: 'Extended Kalman filter based on nonlinear model predictive control', *Ind. Eng. Chem. Res.*, 1994, **33**, pp. 1530–41

6 YANG, T.H., and POLAK, E.: 'Moving horizon control of nonlinear systems with input saturation, disturbances and plant uncertainty', *Int. J. Control*, 1993, **58** (4), pp. 875–903

7 MICHALSKA, H., and MAYNE, D.: 'Robust receding horizon control of constrained nonlinear systems', *IEEE Trans. Autom. Control*, 1993, **38** (11), pp. 1623–33

8 BIEGLER, L.T., and RAWLINGS, J.B.: 'Optimization approaches to nonlinear model predictive control'. Conference on *Chemical process control* (CPC IV), South Padre Island, Texas, 1991, CACHE-AIChE, pp. 543–71

9 MAYNE, D.Q.: 'Nonlinear model predictive control: an assessment'. Fifth international conference on *Chemical process control* (CPC V), Lake Tahoe, CA, 1996

10 KEERTHI, S.S., and GILBERT, E.G.: 'Optimal infinite-horizon feedback laws for a general class of constrained discrete-time systems: stability and moving-horizon approximations', *Journal of Optimisation Theory and Applications*, 1988, pp. 265–93

11 CHEN, H., and ALLGÖWER, F.: 'A quasi-infinite horizon nonlinear model predictive control scheme with guaranteed stability', *Automatica*, 1998, **34** (10), pp. 1205–17

12 NICOLAO, G.D., MAGNI, L., and SCATTOLINI, R.: 'Stability and robustness of nonlinear receding horizon control', in ALLGÖWER, F., and ZHENG, A., (Eds): International symposium on *Nonlinear model predictive control: assessment and future directions*, Ascona, Switzerland, 1998

13 GATTU, G., and ZAFIRIOU, E.: 'Nonlinear quadratic dynamic matrix control with state estimation', *Ind. Eng. Chem. Res.*, 1992, **31** (4), pp. 1096–1104

14 GATTU, G., and ZAFIRIOU, E.: 'Observer based nonlinear quadratic dynamic matrix control for state space and input/output models', *Canadian Journal of Chemical Engineering*, 1995, **73**, pp. 883–95

15 DE OLIVEIRA, S.L., and MORARI, M.: 'Robust model predictive control for nonlinear systems'. Proceedings of 33rd IEEE conference on *Decision and control*, Orlando, FL, 1994, pp. 3561–7

16 DE OLIVEIRA, S.L., NEVISTIC, V., and MORARI, M.: 'Control of nonlinear systems subject to input constraints'. Proceedings of IFAC symposium on *Nonlinear control systems design*, Tahoe City, CA, 1995, pp. 15–20
17 KURTZ, M.J., and HENSON, M.A.: 'Linear model predictive control of input-output linearised processes with constraints'. Fifth international conference on *Chemical process control* (CPC V), Lake Tahoe, CA, 1996
18 ZHENG, A.: 'A computationally efficient nonlinear model predictive control algorithm'. American Control Conference, Albuquerque, NM, 1997
19 CUTLER, C.R., and RAMAKER, B.L.: 'Dynamic matrix control – a computer control algorithm'. Joint Automatic Control Conference, San Francisco, CA, 1980
20 RICKER, N.L., SUBRAHMANIAN, T., and SIM, T.: 'Case studies of model-predictive control in pulp and paper production'. Proceedings of 1988 IFAC workshop on *Model based process control* (Pergamon Press, Oxford, 1989)
21 ZHENG, A., and MORARI, M.: 'Stability of model predictive control with mixed constraints', *IEEE Trans. Autom. Control*, 1995, **40** (10), pp. 1818–23
22 RAWLINGS, J.B., and MUSKE, K.R.: 'The stability of constrained receding horizon control', *IEEE Trans. Autom. Control*, 1993, **38** (10), pp. 1512–6
23 ZHENG, A., BALAKRISHNAN, V., and MORARI, M.: 'Constrained stabilization of discrete-time systems', *International Journal of Robust and Nonlinear Control*, 1995, **5** (5), pp. 461–85
24 MORARI, M., GARCÍA, C.E., LEE, J.H., and PRETT, D.M.: 'Model predictive control' (Prentice-Hall Inc., Englewood Cliffs, N.J, 1997), in preparation
25 LEE, J.H., and COOLEY, B.: 'Robust model predictive control of multi-variable systems using input-output models with stochastic parameters'. American Control Conference, Seattle, WA, 1995
26 ZHENG, A.: 'Nonlinear model predictive control of the Tennessee-Eastman process'. American Control Conference, Philadelphia, PA, 1998
27 MAYNE, D.Q.: 'Nonlinear model predictive control: challenges and opportunities', in ALLGÖWER, F., and ZHENG, A., (Eds): International symposium on the *Assessment and future research directions of nonlinear model predictive control*, Ascona, Switzerland, 1998
28 KOTHARE, M.V., CAMPO, P.J., MORARI, M., and NETT, C.N.: 'A unified framework for the study of anti-windup designs', *Automatica*, 1994, **30** (12), 1869–83. Also presented at the 1993 AIChE Annual Meeting, MO
29 ZHENG, A., KOTHARE, M.V., and MORARI, M.: 'Anti-windup design for internal model control', *International Journal of Control*, 1994, **60** (5), 1015–24. Also presented at the 1993 AIChE Annual Meeting, MO
30 SKOGESTAD, S., and POSTLETHWAITE, I.: 'Multivariable feedback control' (John Wiley & Sons, 1996)
31 DOWNS, J.J., and VOGEL, E.F.: 'A plant-wide industrial-process control problem', *Comput. Chem. Eng.*, 1993, **17** (3), pp. 245–55
32 RICKER, N.L.: 'Decentralized control of the Tennessee Eastman challenge process', *Journal of Process Control*, 1996, **6**, 205–21
33 MORARI, M., and RICKER, N.L.: 'Predictive methods toolbox for Matlab (test version)', 1993

34 ZHENG, A.: 'Some practical issues and possible solutions for nonlinear model predictive control', in ALLGÖWER, F., and ZHENG, A., (Eds): 'Nonlinear model predictive control–current status and future research directions, progress in systems and control theory' (Birkhauser, 1999)

Chapter 9

Long-prediction-horizon nonlinear model predictive control

Masoud Soroush and Hossein M. Soroush

Abstract

One of the advantages of long-prediction-horizon model predictive control (MPC) is its applicability to processes with nonminimum-phase behaviour. Motivated by this attractive feature of MPC, a long-prediction-horizon MPC formulation is used to derive an approximate input-output-linearising nonlinear control law for hyperbolically stable, single-input single-output processes, whether nonminimum-phase or minimum phase. Indeed, the problem of nonlinear control of a class of nonminimum-phase processes is solved by exploiting further the connections between model predictive control and input-output linearisation. The derived control law has one single tunable parameter, and thus is very easy to tune. It is applied to linear processes, and the resulting linear control law is presented.

9.1 Introduction

Model predictive control is a very broad controller-synthesis methodology. The considerable broadness comes from the optimisation-based nature of MPC and the great number of tunable parameters that a model predictive controller can possess. Its optimisation-based nature allows one to choose any desirable performance index irrespective of the type of process model available. Its numerous tunable parameters provide many degrees of freedom to obtain a desirable closed-loop response. The performance of a model predictive controller can be as nonrobust and aggressive as that of a deadbeat controller, or as robust and slow as that of a steady-state controller (i.e. a model-based controller that takes the process output to its setpoint, with a constant manipulated input when the process disturbances and setpoint are constant).

The literature on model predictive control has been mainly on practical issues in the implementation of MPC, controller tuning and stability analysis [1, 2]. Several recent studies have established the connections between MPC and existing analytical control methods [3–6]. A few special cases of unconstrained single-input single-output (SISO) MPC are listed in Table 9.1. As the prediction horizon is increased and the control horizon is decreased, a more robust but less aggressive controller is obtained. In Reference 3 we focused on shortest-prediction-horizon MPC [MPC with prediction horizons equal to relative orders (degrees)] and showed that in special cases the shortest-prediction-horizon MPC leads to the following:

- input-output linearising control laws that inherently include optimal windup and directionality compensators [7]
- model state feedback control (MSFC) [8] and modified internal model control (IMC) [9] laws that inherently include optimal directionality compensators
- proportional-integral (PI) and proportional-integral-derivative (PID) controllers that inherently include optimal windup and directionality compensators.

In the framework of model-predictive control, it is well known that large prediction horizons should be used for nonminimum-phase processes. For example, Hernandez and Arkun [10] developed a p-inverse (long prediction horizon) control law for stable, SISO, nonlinear, discrete-time processes with arbitrary relative order. Motivated by their results, in this chapter we derive a nonlinear feedback control law for stable, SISO, nonlinear, discrete-time processes, whether nonminimum-phase or minimum-phase. The derived control law is basically a discrete-time long-prediction-horizon model predictive control law with the shortest control horizon and can be considered as an approximate, input-output, linearising control law [11]. Therefore, this study addresses nonlinear control of nonminimum-phase and minimum-phase processes by exploiting further the connections between model predictive control and input-output linearisation [3]. The control law is the discrete-time analogue of one of those developed for continuous-time processes in References 12 and 13. This method of controlling

Table 9.1 Unconstrained SISO model predictive controllers with no penalty on manipulated input magnitude or rate of change (r = relative order) [3, 10]

Prediction horizon	*Control horizon*	*Reference trajectory*	*Resulting controller*
r	1	No	Deadbeat
r	1	Yes	I-O linearisation
r	1	Yes	Modified IMC [9]
r	1	Yes	MSFC
$r = 1$	1	Yes	PI, PID

nonminimum-phase processes does not possess the limitations of the existing methods described in References 14–16.

9.2 Scope and preliminaries

We consider SISO nonlinear processes of the form:

$$\begin{aligned} \bar{x}(k+1) &= \Phi[\bar{x}(k), u(k)], \quad \bar{x}(0) = \bar{x}_0 \\ \bar{y}(k) &= h[\bar{x}(k)] + d \end{aligned} \tag{1}$$

where $\bar{x} \in X \subset \mathbf{R}^n$ denotes the vector of process state variables, $u \in U \subset \mathbf{R}$ is the manipulated input, $\bar{y} \in h(X) \subset \mathbf{R}$ is the process output, and $d \in D \subset \mathbf{R}$ is an unmeasurable constant disturbance. Here X and U are open connected sets and D is a connected set.

We make the following assumptions: (a) the nominal equilibrium pair of the process (x_{ss}, u_{ss}) is hyperbolically stable; that is, all eigenvalues of the Jacobian of the process evaluated at (x_{ss}, u_{ss}) lie strictly inside the unit circle; (b) $\Phi(x, u)$ and $h(x)$ are smooth vector functions on $X \times U$ and X, respectively; and (c) for a process of the form of (1), a discrete-time model of the following form is available:

$$\begin{aligned} x(k+1) &= \Phi[x(k), u(k)], \quad x(0) = x_0 \\ y(k) &= h[x(k)] + d \end{aligned} \tag{2}$$

where $x \in X \subset \mathbf{R}^n$ denotes the vector of model state variables and $y \in h(X) \subset \mathbf{R}$ is the model output. Relative order (degree) of the controlled output y with respect to the manipulated input u is denoted by r, where the relative order r is the smallest integer for which $y(k+r)$ depends explicitly on the manipulated input $u(k)$.

9.3 Optimisation problem: model predictive control law

We consider a moving-horizon minimisation problem of the special form

$$\min_{\bar{U}(k)} [y_d(k+P) - \hat{y}(k+P)]^2 \tag{3}$$

subject to:

$$\begin{aligned} x(k+1) &= \Phi[x(k), u(k)] \\ y(k) &= h[x(k)] \\ u(k+l) &= u(k+l-1), l = M, \ldots, P-r \end{aligned}$$

where: $P \geq r$ is the prediction horizon; M is the control horizon; $\bar{U}(k) =$

$[u(k)\dots u(k+M-1)]^T$; y_d is a reference trajectory, given by

$$y_d(k+P) = (1+\beta)^P y_{sp} - \binom{P}{P}\beta^P y_d(k) - \cdots - \binom{P}{1}\beta y_d(k+P-1)$$

with $y_d(k+l) = \hat{y}(k+l), l = 0, \dots, P-1$; β is a tunable scalar parameter chosen such that $|\beta| < 1$; $y_{sp} \in Y \subset \mathbf{R}$ is the output setpoint; Y is the set of all y_{sp} for which there exists a $u_{ss} \in U$ and an $x_{ss} \in U$ satisfying $x_{ss} = \Phi(x_{ss}, u_{ss})$ and $y_{sp} = h(x_{ss}) + d$, for every $d \in D$;

$$\binom{a}{b} = \frac{a!}{b!(a-b)!}$$

$\hat{y}$ is the predicted future value of the controlled output, given by:

$$\begin{aligned}
&\hat{y}(k+l) \equiv \bar{y}(k) + h^l[x(k)] - h[x(k)], \quad l = 0, \dots, r-1 \\
&\hat{y}(k+r) \equiv \bar{y}(k) + h^r[x(k), u(k)] - h[x(k)] \\
&\hat{y}(k+l) \equiv \bar{y}(k) + h^l[x(k), u(k), \dots, u(k+l-r)] - h[x(k)], l = r+1, \dots, P
\end{aligned} \tag{4}$$

and:

$$\begin{aligned}
h^0[x(k)] &\equiv h[x(k)] \\
h^l[x(k)] &\equiv h^{l-1}[\Phi(x(k), u(k))], l = 1, \dots, r-1 \\
h^r[x(k), u(k)] &\equiv h^{r-1}[\Phi(x(k), u(k))] \\
h^{r+1}[x(k), u(k), u(k+1)] &\equiv h^r[\Phi(x(k), u(k)), u(k+1)] \\
&\vdots \\
h^P[x(k), u(k), \dots, u(k+P-r)] &\equiv h^{P-1}[\Phi(x(k), u(k)), u(k+1), \dots, u(k+P-r)].
\end{aligned}$$

9.4 Nonlinear feedforward/state feedback design

Solving the moving-horizon minimisation problem of (3) leads to a feedforward/state feedback. We obtain a global solution to the minimisation problem with the control horizon $M = 1$ by making the following two assumptions:

1. For every $\bar{x} \in X$, every $y_{sp} \in Y$ and every $d \in D$, the algebraic equation

$$\phi_P(\bar{x}, u) = (\beta+1)^P (y_{sp} - d) \tag{5}$$

where

$$\begin{aligned}\phi_P[\bar{x}(k),u(k)] \equiv \binom{P}{P}\beta^P h[\bar{x}(k)] + \cdots \\ + \binom{P}{P-r+1}\beta^{P-r+1}h^{r-1}[\bar{x}(k),u(k)] \\ + \binom{P}{P-r}\beta^{P-r}h^{r}[\bar{x}(k),u(k)] \\ + \binom{P}{P-r-1}\beta^{P-r-1}h^{r+1}[\bar{x}(k),u(k),u(k)] + \cdots \\ + \binom{P}{0}h^{P}[\bar{x}(k),u(k),\ldots,u(k)]\end{aligned} \tag{6}$$

has a real root inside U for u.

2. For every $\bar{x} \in X$ and every $u \in U$

$$\frac{\partial \phi_P(x,u)}{\partial u} \neq 0 \tag{7}$$

The corresponding feedforward/state feedback that satisfies (5) – that is, a global minimiser of (3) – is denoted by

$$u = \Psi_P(\bar{x}, y_{sp} - d). \tag{8}$$

Note that the preceding feedforward/state feedback was obtained by setting the control horizon $M = 1$. When $P \to \infty$, the feedforward/state feedback of (8) is simply a solution for u of

$$\bar{x}_{ss} = \Phi[\bar{x}_{ss}, u], \quad y_{sp} - d = h[\bar{x}_{ss}]$$

which is just a steady-state (feedforward) controller (with no feedback feature).

Theorem 1: For a process of the form of (1), the closed-loop system under the feedforward/state feedback of (8) is asymptotically stable if the following conditions hold:

(a) The equilibrium pair of the process, $(\bar{x}_{ss}, u_{ss})$, corresponding to $y_{ss} = [y_{sp} - d]$, is hyperbolically stable.
(b) The tunable parameter P is chosen to be sufficiently large.
(c) The tunable parameter β is chosen such that all the eigenvalues of $[\frac{\partial}{\partial \bar{x}}\Phi(\bar{x},u)|_{(\bar{x}_{ss},u_{ss})} + \beta I]$ lie inside the unit circle.

Furthermore, as $P \to \infty$ the feedforward/state feedback places the n eigenvalues of the Jacobian of the closed-loop system evaluated the equilibrium pair $(\bar{x}_{ss}, u_{ss})$ at the n eigenvalues of the Jacobian of the open-loop process evaluated the same equilibrium pair.

The proof is given in the Appendix (Section 9.10.1). The optimising feedforward/state feedback of (8) tries to force the process output to match the linear reference trajectory, y_d. In other words, it induces *approximately* the offset-free, linear, input-output, closed-loop response

$$(q+\beta)^P \bar{y}(k) = (1+\beta)^P y_{sp} \tag{9}$$

where q is the forward shift operator, or equivalently

$$\beta^P \binom{P}{P} \bar{y}(k) + \beta^{P-1} \binom{P}{P-1} \bar{y}(k+1) + \cdots + \binom{P}{0} \bar{y}(k+P) = (1+\beta)^P y_{sp} \tag{10}$$

Therefore, the feedforward/state feedback of (8) is an approximate input-output linearising feedforward/state feedback. The single tunable parameter determines the speed of the closed-loop process output response; that is, the smaller the value of $|\beta|$, the faster the response. Setting $\beta = 0$ in the feedforward/state feedback of (8) results in the p-inverse (long prediction horizon) control law of Hernandez and Arkun [10].

9.5 Nonlinear feedback controller design

In order for the feedforward/state feedback of (8) to induce an offset-free closed-loop process output response, the unmeasurable disturbance should be estimated from the available measurements. An estimate of the disturbance is $\hat{d} = \bar{y} - \tilde{y}$, where $\tilde{y}$ is an estimate of the 'disturbance-free' controlled output. One way of estimating the 'disturbance-free' controlled output, $\tilde{y}$, is to simply use the model of (2), as in IMC [9]; that is, $\tilde{y} = h(x)$ and $\hat{d} = \bar{y} - h(x)$. The estimation of d and the use of the estimate along with the feedforward/state feedback of (8) lead to a feedback control law with integral action, as stated in the following theorem.

Theorem 2: For a process of the form of (1), the closed-loop system under the error-feedback control law:

$$\begin{aligned} x(k+1) &= \Phi\{x(k), \Psi_P[x(k), e(k) + h(x(k))]\}, x(0) = x_0 \\ u(k) &= \Psi_P[x(k), e(k) + h(x(k))] \end{aligned} \tag{11}$$

where $e = y_{sp} - \bar{y}$, is asymptotically stable if the following conditions hold:

(a) The equilibrium pair of the process, (x_{ss}, u_{ss}), corresponding to $y_{ss} = [y_{sp} - d]$ is hyperbolically stable.

(b) The tunable parameter P is chosen to be sufficiently large.
(c) The tunable parameter β is chosen such that all the eigenvalues of $[\frac{\partial}{\partial x}\Phi(x,u)|_{(x_{ss},u_{ss})} + \beta I]$ lie inside the unit circle.

Furthermore, the error-feedback control law of (11) has integral action.

The proof is given in the Appendix (Section 9.10.2). A block diagram of the error-feedback control law of (11) is depicted in Figure 9.1. The integral action of the control law of Theorem 2 ensures offset-free closed-loop process output response: (i) in the presence of the unmeasurable disturbance; and (ii) in the presence of constant process-model mismatch as long as the nominal asymptotic stability of the closed-loop system is preserved.

9.6 Application to linear processes

Consider the class of time-invariant, linear processes with a model of the form:

$$\begin{aligned} x(k+1) &= Ax(k) + bu(k), \quad x(0) = x_0 \\ y(k) &= cx(k) \end{aligned} \tag{12}$$

where A, b and c are $n \times n$, $n \times 1$ and $1 \times n$ constant matrices, respectively. This class of processes is a special case of (1).

For this class of processes the function $\phi_P(x,u)$ is linear in u,

$$\phi_P(\bar{x},u) = \sum_{l=0}^{P} \binom{P}{P-l} \beta^{P-l} cA^l \bar{x} + \sum_{l=r}^{P} \sum_{j=r}^{l} \binom{P}{P-l} \beta^{P-l} cA^{j-1} bu$$

and thus it can be solved for u analytically:

$$u = \Psi_P(\bar{x}, y_{sp} - d) = \frac{(1+\beta)^P (y_{sp} - d) - \sum_{l=0}^{P} \binom{P}{P-l} \beta^{P-l} cA^l \bar{x}}{\sum_{l=r}^{P} \sum_{j=r}^{l} \binom{P}{P-l} \beta^{P-l} cA^{j-1} b} \tag{13}$$

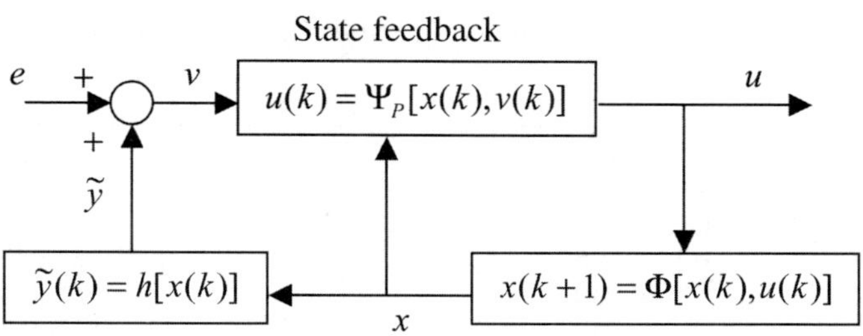

Figure 9.1 Error-feedback control structure

Application of the control law of (11) to processes of the form of (12) leads to the following linear control law:

$$x(k+1) = Ax(k) + bu(k)$$
$$u(k) = \frac{(1+\beta)^P[e(k) + cx(k)] - \sum_{l=0}^{P}\binom{P}{P-l}\beta^{P-l}cA^l x(k)}{\sum_{l=r}^{P}\sum_{j=r}^{l}\binom{P}{P-l}\beta^{P-l}cA^{j-1}b} \tag{14}$$

As stated in Theorem 2, in the limit that $P \to \infty$ the preceding control law places the n eigenvalues of the linear closed-loop system at the n eigenvalues of the open-loop process. This is illustrated by the linear example given below.

Example 1: Consider the linear process

$$A = \begin{bmatrix} 0 & 1 \\ -0.125 & 0.75 \end{bmatrix}, \quad b = \begin{bmatrix} 0 \\ 1 \end{bmatrix}, \quad c = [2 \quad -1]$$

which has the transfer function

$$G(z) = \frac{2-z}{(z-0.5)(z-0.25)}$$

It is nonminimum-phase (has an outside-the-unit-circle zero at $z = 2$) and asymptotically stable (has two inside-the-unit-circle poles at $z = 0.5$ and $z = 0.25$). For $\beta = 0.0, -0.1, -0.3$, both eigenvalues of $(A + \beta I)$ lie inside the unit circle. The Jacobian of the closed-loop system under the feedforward/state feedback of (13) for processes of the form of (12) is

$$Jcl_P = A - \frac{bc[A+\beta I]^P}{\sum_{l=r}^{P}\sum_{j=r}^{l}\binom{P}{P-l}\beta^{P-l}cA^{j-1}b}$$

The closed-loop eigenvalues, $\lambda_1(Jcl_P)$ and $\lambda_2(Jcl_P)$, for several values of P are

Table 9.2 Closed-loop eigenvalues of Example 1 for several values of P and β

P	$\beta = 0.0$	$\beta = -0.1$	$\beta = -0.3$
1	0.00, 2.00	0.10, 2.00	0.30, 2.00
2	0.14, −3.64	0.19, −1.22	$0.24 \pm 0.07i$
3	$0.13 \pm 0.08i$	$0.23 \pm 0.07i$	0.25, 0.44
4	$0.29 \pm 0.05i$	0.25, 0.39	0.25, 0.48
5	0.25, 0.42	0.25, 0.46	0.25, 0.50

given in Table 9.2; the closed-loop eigenvalues converge to the open-loop eigenvalues $z = 0.5$ and $z = 0.25$ as $P \to \infty$.

9.7 Conclusions

The very broad controller-synthesis methodology of model predictive control was used to solve the problem of nonlinear control of a class of nonminimum-phase processes. In particular, MPC with large prediction horizon and the shortest control horizon was used to derive a nonlinear feedback control law for unconstrained, hyperbolically stable, SISO, nonlinear, discrete-time systems, whether nonminimum-phase or minimum-phase. The control law can be considered as an approximate input-output linearising controller. It ensures asymptotic tracking of the setpoint in the presence of constant process-model mismatch and unmeasurable disturbances as long as the nominal closed-loop asymptotic stability is preserved. There are no limitations on the order, relative order, or number of unstable modes of the zero dynamics of processes to which the control law is applicable.

9.8 Acknowledgments

Financial support from the National Science Foundation through grant CTS-9703278, DuPont, and Fluor Daniel, Inc., is gratefully acknowledged.

9.9 References

1 MORARI, M., and LEE, J.: 'Model predictive control: past, present and future', *Comput. Chem. Eng.*, 1999, **23**, pp. 667–82
2 MUSKE, K.R., and RAWLINGS, J.B.: 'Model predictive control with linear models', *AIChE J.*, 1993, **39**, pp. 262–87
3 SOROUSH, M., and MUSKE, K.: 'Analytical model predictive control', in ALLGÖWER, F., and ZHENG, A. (Eds.): 'Nonlinear model predictive control', Progress in Systems and Control Theory series, Vol. 26 (Birkhauser-Verlag, Basel, 2000)
4 SOROUSH, M., and SOROUSH, H.M.: 'Input-output linearizing nonlinear model predictive control', *Int. J. Control.*, 1997, **68** (6), pp. 1449–73
5 SOROUSH, M., and KRAVARIS, C.: 'Discrete-time nonlinear controller synthesis by input/output linearisation', *AIChE J.*, 1992, **38** (12), pp. 1923–45
6 VALLURI, S., SOROUSH, M., and NIKRAVESH, M.: 'Shortest-prediction-horizon nonlinear model predictive control', *Chem. Eng. Sci.*, 1998, **53** (2), pp. 273–92
7 SOROUSH, M., and VALLURI, S.: 'Optimal directionality compensation in processes with input saturation nonlinearities', *Int. J. Control*, 1999, **72** (17), pp. 1555–64

8 COULIBALY, E., MAITI, S., and BROSILOW, C.: 'Internal model predictive control (IMPC)', *Automatica*, 1995, **31**, pp. 1471–82
9 ZHENG, A., KOTHARE, M.V., and MORARI, M.: 'Anti-windup design for internal model control', *Int. J. Control*, 1994, **60**, pp. 1015
10 HERNANDEZ, E., and ARKUN, Y.: 'Study of the control-relevant properties of back-propagation neural network models of nonlinear dynamical systems', *Comput. Chem. Eng.*, 1992, **16**, pp. 227–40
11 ALLGÖWER, F., and DOYLE III, F.J.: 'Approximate I/O-linearisation of nonlinear systems', in BERBER, R., and KRAVARIS, C. (Eds.): 'Nonlinear model-based process control', NATO ASI Series (Kluwer, Dordrecht, 1998), pp. 235–74
12 KANTER, J., SOROUSH, M., and SEIDER, W.: 'Continuous-time, nonlinear feedback control of stable processes', *Ind. Eng. Chem. Research*, 2001, **40** (9), pp. 2069–78
13 KANTER, J., SOROUSH, M., and SEIDER, W.: 'Continuous-time nonlinear control of stable non-minimum-phase processes'. IFAC ADCHEM Conference, Italy, 2000, pp. 407–12
14 KRAVARIS, C., and DAOUTIDIS, P.: 'Nonlinear state feedback control of second-order non-minimum phase nonlinear systems', *Comput. Chem. Eng.*, 1990, **14**, pp. 439–49
15 WRIGHT, R.A., and KRAVARIS, C.: 'Non-minimum-phase compensation for nonlinear processes', *AIChE J.*, 1992, **38**, pp. 26–40
16 DOYLE III, F.J., ALLGÖWER, F., and MORARI, M.: 'Normal form approach to approximate input-output linearisation for maximum phase nonlinear SISO systems', *IEEE Trans. Autom. Control*, 1996, **41**, pp. 305–9

9.10 Appendix

9.10.1 Proof of Theorem 1

The Jacobian of the closed-loop system under the feedforward/state feedback is

$$Jcl_p(\bar{x}, u) = \frac{\partial \Phi(\bar{x}, u)}{\partial \bar{x}} + \frac{\partial \Phi(\bar{x}, u)}{\partial u} \frac{\partial \Psi_P(\bar{x}, y_{sp} - d)}{\partial \bar{x}} \tag{15}$$

We prove that under the conditions of Theorem 1 as $P \to \infty$, $Jcl_p(\bar{x}_{ss}, u_{ss}) \to J_{ol}(\bar{x}_{ss}, u_{ss})$.

Since $\phi_P(\bar{x}, u) = y_{sp} - d$ and $u = \Psi_P(\bar{x}, y_{sp} - d)$, the partial derivative of $\phi_P(\bar{x}, u) = y_{sp} - d$ with respect to $\bar{x}$ yields

$$\frac{\partial \phi_P(\bar{x}, u)}{\partial \bar{x}} + \frac{\partial \phi_P(\bar{x}, u)}{\partial u} \frac{\partial \Psi_P(\bar{x}, y_{sp} - d)}{\partial \bar{x}} = [0 \ldots 0] \tag{16}$$

Using the definition of $\phi_P(\bar{x}, u)$ given by (6), we obtain

$$\frac{\partial\phi_P(\bar{x},u)}{\partial u}\frac{\partial\Psi_P(\bar{x},y_{sp}-d)}{\partial\bar{x}} = -\frac{\partial\phi_P(\bar{x},u)}{\partial\bar{x}} = -\binom{P}{P}\beta^P\frac{\partial h(\bar{x})}{\partial\bar{x}} - \cdots$$
$$-\binom{P}{P-r+1}\beta^{P-r+1}\frac{\partial h^{r-1}(\bar{x})}{\partial\bar{x}} - \binom{P}{P-r}\beta^{P-r}\frac{\partial h^{r}(\bar{x},u)}{\partial\bar{x}}$$
$$-\binom{P}{P-r-1}\beta^{P-r-1}\frac{\partial h^{r+1}(\bar{x},u,u)}{\partial\bar{x}} - \cdots - \binom{P}{0}\frac{\partial h^{P}(\bar{x},u,\ldots,u)}{\partial\bar{x}}$$

Using the definition of $h^1, \ldots, h^P$, we can write

$$\begin{aligned}
\frac{\partial h^1(\bar{x})}{\partial\bar{x}} &= \frac{\partial h(\bar{x})}{\partial\bar{x}}\frac{\partial\Phi(\bar{x},u)}{\partial\bar{x}} = \frac{\partial h(\bar{x})}{\partial\bar{x}}J_{ol}\\
\frac{\partial h^2(\bar{x})}{\partial\bar{x}} &= \frac{\partial h^1(\bar{x})}{\partial\bar{x}}\frac{\partial\Phi(\bar{x},u)}{\partial\bar{x}} = \frac{\partial h(\bar{x})}{\partial\bar{x}}J_{ol}^2\\
&\vdots\\
\frac{\partial h^P(\bar{x},u,u,\ldots,u)}{\partial\bar{x}} &= \frac{\partial h^{P-1}(\bar{x})}{\partial\bar{x}}\frac{\partial\Phi(\bar{x},u)}{\partial\bar{x}} = \frac{\partial h(\bar{x})}{\partial\bar{x}}J_{ol}^P
\end{aligned} \tag{17}$$

Therefore,

$$\frac{\partial\phi_P(\bar{x},u)}{\partial u}\frac{\partial\Psi_P(\bar{x},y_{sp}-d)}{\partial\bar{x}} = -\frac{\partial h(\bar{x})}{\partial\bar{x}}[J_{ol}+\beta I]^P \tag{18}$$

Since the equilibrium pair $(\bar{x}_{ss}, u_{ss})$ is asymptotically stable in open-loop (eigenvalues of $J_{ol}(\bar{x}_{ss}, u_{ss})$ are all inside the unit circle) and we choose β such that all eigenvalues of ($J_{ol}(\bar{x}_{ss}, u_{ss}) + \beta I$) are inside the unit circle,

$$\lim_{P\to\infty}[J_{ol}(\bar{x}_{ss},u_{ss})+\beta I]^P = 0$$

and thus by (18)

$$\lim_{P\to\infty}\left[\frac{\partial\phi_P(\bar{x},u)}{\partial u}\right]_{(\bar{x}_{ss},u_{ss})}\lim_{P\to\infty}\left[\frac{\partial\Psi_P(\bar{x},y_{sp}-d)}{\partial\bar{x}}\right]_{(\bar{x}_{ss},u_{ss})} = 0 \tag{19}$$

Because

$$\left[\frac{\partial\phi_P(\bar{x},u)}{\partial u}\right]_{(\bar{x}_{ss},u_{ss})}$$

is nonzero (see the condition of (7)),

$$\lim_{P\to\infty}\left[\frac{\partial\Psi_P(\bar{x},y_{sp}-d)}{\partial\bar{x}}\right]_{(\bar{x}_{ss},u_{ss})} = 0 \tag{20}$$

This, together with (15) proves that

$$\lim_{P\to\infty} Jcl_p(\bar{x}_{ss}, u_{ss}) = \left[\frac{\partial\phi(\bar{x}, u)}{\partial\bar{x}}\right]_{(\bar{x}_{ss}, \bar{u}_{ss})} = J_{ol}(\bar{x}_{ss}, u_{ss}) \tag{21}$$

9.10.2 Proof of Theorem 2

The closed-loop system under the control law of (11) is given by

$$\begin{aligned} \bar{x}(k+1) &= \Phi(\bar{x}(k), \Psi_P[x(k), y_{sp} - d + h(x(k)) - h(\bar{x}(k))]) \\ x(k+1) &= \Phi(x(k), \Psi_P[x(k), y_{sp} - d + h(x(k)) - h(\bar{x}(k))]) \end{aligned} \tag{22}$$

and its Jacobian at the equilibrium point (x_{ss}, u_{ss}) by

$$Jcl_p(x_{ss}, u_{ss}) = \begin{bmatrix} J_{ol}(x_{ss}, u_{ss}) - \Theta & Q + \Theta \\ -\Theta & J_{ol}(x_{ss}, u_{ss}) + Q + \Theta \end{bmatrix}$$

where

$$Q = \left[\frac{\partial\Phi(\bar{x}, u)}{\partial u} \frac{\partial\Psi_P(\bar{x}, v)}{\partial\bar{x}}\right]_{(x_{ss}, u_{ss})}, \quad \Theta = \left[\frac{\partial\Phi(\bar{x}, u)}{\partial u} \frac{\partial\Psi_P(\bar{x}, v)}{\partial v} \frac{\partial h(x)}{\partial x}\right]_{(x_{ss}, u_{ss})}$$

Since the above closed-loop Jacobian and the matrix

$$\begin{bmatrix} J_{ol}(x_{ss}, u_{ss}) & 0 \\ -\Theta & J_{ol}(x_{ss}, u_{ss}) + Q \end{bmatrix} \tag{23}$$

are similar, the eigenvalues of $Jcl_p(x_{ss}, u_{ss})$ are the same as the eigenvalues of the matrix of (23). The similarity transformation is

$$\begin{bmatrix} I & -I \\ 0 & I \end{bmatrix}$$

Since the matrix of (23) is lower block triangular, its eigenvalues are the eigenvalues of 1,1 and 2,2 blocks. The 1,1 block has eigenvalues inside the unit circle since these are the open-loop eigenvalues. The eigenvalues of the 2,2 block are the eigenvalues of the process under the feedforward/state feedback of (8). In Theorem 1 we proved that the eigenvalues of the 2,2 block lie inside the unit circle when conditions (a)–(c) hold.

To prove that the control law of (11) induces an offset-free response (i.e. $e_{ss} = 0$), consider the closed-loop system at steady state:

$$\begin{aligned} \bar{x}_{ss} &= \Phi\{\bar{x}_{ss}, \Psi_P[x_{ss}, e_{ss} + h(x_{ss})]\} \\ x_{ss} &= \Phi\{x_{ss}, \Psi_P[x_{ss}, e_{ss} + h(x_{ss})]\} \\ u_{ss} &= \Psi_P[x_{ss}, e_s + h(x_{ss})] \end{aligned} \tag{24}$$

with $\bar{x}_{ss} = x_{ss}$. According to the definition of $\phi_P(\cdot, \cdot)$,

$$\phi_P(\bar{x}_{ss}, u_{ss}) = (1 + \beta)^P h(x_{ss})$$

and according to the definition of $\Psi_P(\cdot, \cdot)$, the third equation in (24) implies that

$$\phi_P(\bar{x}_{ss}, u_{ss}) = (1 + \beta)^P [e_{ss} + h(x_s)]$$

Thus, $e_{ss} = 0$.

Part IV

Chapter 10

Nonlinear control of industrial processes

Babatunde A. Ogunnaike

Abstract

As a result of increased customer demand for consistent attainment of high product quality, coupled with increasingly stringent safety and environmental regulations, and intensified global competition, the current drive in the chemical and allied industries has been towards more efficient utilisation of existing assets (especially capacity and energy) rather than new capital expenditure. The result is that a growing number of industrial processes must now operate under conditions that emphasise their inherent nonlinearities. Nonlinear control is thus becoming more important in industrial practice. This chapter assesses the current status of nonlinear control applications in the chemical industry, discusses some of the most pertinent issues of, and barriers to, practical implementation, and presents an actual industrial application to illustrate the main points.

10.1 Introduction

It is well known that virtually all processes of practical importance exhibit some degree of nonlinear behaviour. Nevertheless, the vast majority of well-established controller design techniques are for linear systems. Such techniques typically work well in practice for processes that exhibit only mildly nonlinear dynamic behaviour. More recently, increasingly stringent requirements on product quality and energy utilisation, as well as on safety and environmental responsibility, demand that a growing number of industrial processes operate in such a manner as to emphasise their inherent nonlinearity even more. There is therefore increased industrial and academic interest in the development and implementation of controllers that will be effective when process nonlinearities cannot be ignored without serious consequences.

The growing interest of the process control community in nonlinear control is reflected in several reviews of currently used techniques (see, for example, References 1–4). To be sure, many significant theoretical and practical issues remain unresolved; nevertheless, the impact of the available theory on industrial practice is becoming more noticeable. First, observe that it has become standard industrial practice to use certain simple nonlinear elements to improve performance in some control loops – for example, square root correction in flow control (see Reference 5). But beyond such simple applications, there is a growing number of more complex nonlinear control applications that have appeared in the open literature – for example, see Reference 6, model based control of an industrial extruder; Reference 7, generic model control of an industrial blast furnace; Reference 8, geometric nonlinear model-based control of a binary distillation column; Reference 9, geometric nonlinear control of an industrial CO_2 adsorption/desorption pilot plant process; References 10 and 11, nonlinear control of industrial pH processes; Reference 12, nonlinear model predictive control of an industrial packed-bed reactor; Reference 13, nonlinear model predictive control for economic optimisation and control of gas processing plants. For a more recent overview of nonlinear model predictive control applications, see Reference 14.

However, while the number of industrial applications of nonlinear control is growing, a careful consideration of the current opportunities *vis-à-vis* the currently available theory indicates that such applications are, in fact, not as widespread as they could be. This chapter has a twofold overall objective:

1. to discuss the issues involved in implementing nonlinear control in industry: assessing the current status (the problems and challenges) and identifying the means by which the impact of nonlinear control on industrial practice can be improved
2. to use an industrial case study (a) to demonstrate the potential impact of nonlinear control, appropriately applied; and (b) to illustrate the main issues involved in successful industrial implementations of nonlinear control.

10.2 Applying nonlinear control to industrial processes

A significant proportion of the demands placed on the typical industrial production facility translates into one, or more, of the following:

1. the need to increase capacity (to meet overall market demands)
2. the need to improve product quality (to meet individual customer demands)
3. the need to reduce environmental emissions (to meet safety and environmental regulatory demands).

Traditionally, it has been customary to adopt the 'capital expenditure' approach in solving these problems: for example, building new production facilities to handle the 'capacity problem'; adding blending facilities to handle the 'quality problem' and redesigning and retrofitting processing units to handle the 'environmental problem'. More recently, however, increasing global competition has dictated the current trend towards finding alternative solutions requiring little or no capital expenditure. This almost invariably implies seeking effective control solutions first, wherever possible. But when most processes are operated under the conditions dictated by these stringent market, customer and environmental demands, the tendency is for the inherent process nonlinearities to become more pronounced – making it more difficult to obtain acceptable solutions from traditional linear controller design techniques. The prevailing global economic conditions thus continue to create opportunities for the application of nonlinear control techniques.

Given the current potential for nonlinear control to contribute significantly to industrial productivity, we now consider the issues that must be addressed for such potential to be realised fully.

10.2.1 Quantitative needs assessment

It is widely accepted that only about 10–20 per cent of industrial control problems require the application of so-called 'advanced control'. It is also accepted that processes in which such problems are encountered account for close to 80 per cent of the revenue. Of the industrial control problems in need of advanced control applications, there is now an increasing realisation that a certain proportion cannot be solved effectively by linear techniques, which constitute the bulk of the most widely applied of these advanced techniques. However, the application of nonlinear techniques requires incrementally greater investments in implementation effort and costs, and such costs must therefore be economically justifiable. Thus, being able to answer the following questions as objectively as possible will increase the impact of nonlinear control in industrial practice:

1. For which problem is the application of nonlinear control critical to the achievement of the desired operational objectives (and which of the available tools is most appropriate for the specific application)?
2. How does the cost of implementation compare to the potential benefits to be derived from the application?

For many of the documented applications of nonlinear control, these questions were relatively straightforward to answer. When the process nonlinearity is obvious, and severe enough (as with the application soon to be discussed), the need for nonlinear control is usually clear. By the same token, if a critical process is virtually inoperable with linear controllers, it will be relatively straightforward to quantify the benefit of nonlinear control. The vast, virtually untapped – and currently difficult to quantify – potential for nonlinear control lies with the class of problems for which linear control methods are applicable, but for which nonlinear methods will result in significant process performance improvements. In this regard, observe that theoretical tools for quantifying the degree of process interaction (and process conditioning) have been useful in assessing the applicability of multivariable control and have thereby promoted industrial application. Similar tools for measuring the degree of process nonlinearity could conceivably play a commensurate role in promoting the industrial application of nonlinear control methods.

10.2.2 Technological and implementation issues

There are a few major factors that currently prevent the widespread use of nonlinear control, even in the cases where the need is obvious, and the potential benefit is known to be substantial:

1. *Control technology*: The typical analytical tools required for rigorous nonlinear systems analysis and controller design still remain largely inaccessible to all but a few researchers concerned with such problems. Naturally, these techniques tend to be more complicated and less transparent and 'intuitive' than the linear techniques.
2. *Model development*: Virtually all high performance controllers are model based; and nonlinear controllers in general require nonlinear process models. Developing linear process models can be difficult enough in practice; developing nonlinear models is several orders of magnitude more difficult.
3. *Implementation*: Most nonlinear controller design techniques give rise to complex controllers that often require unique, specialised software and hardware resources for real-time implementation.

These issues arise primarily because of the intrinsic characteristics of nonlinear systems. First, because nonlinearity is an intrinsically more complex phenomenon to analyse than linearity, nonlinear systems are understandably more difficult to analyse, and nonlinear controllers more difficult to design; by extension, nonlinear control technology will therefore not be as widely accessible as its linear counterpart.

Second, because of all the nice properties enjoyed by linear systems (additivity, homogeneity, superposition, *etc.*) linear model development is relatively straightforward, in concept, if sometimes tedious in practice. The literature on linear model identification from empirical plant data in particular, is vast, and essentially complete; and industrial practice of linear empirical modelling is reasonably well developed. When the desired process model is to be nonlinear, however, many additional issues immediately arise by virtue of this departure from linearity, the most important of which has to do with what modelling approach to adopt: the theoretical (or first-principles) approach, the empirical approach or the 'hybrid' approach.

The first-principles approach is often not employed because it requires a significant amount of process knowledge which may not always be available; when such knowledge is available, the resulting model may simply be too difficult to be useful for controller design purposes. The empirical approach has the advantage of depending strictly on data, but it requires an *a priori* choice of model structure (itself a very difficult task); in addition it requires a very careful design of the input sequence to be used for the identification (see, for example, Reference 15). An increasingly promising approach is the so-called 'grey-box' or hybrid approach in which basic first-principles information is augmented with empirical data, thereby taking advantage of the benefits of each approach. For some sample hybrid modelling applications, see, for example, References 16–19.

Finally, by definition, and intrinsically, nonlinear systems tend to defy classification: they are all characterised by the property they lack – linearity. Each nonlinear control application thus tends to be unique and specialised, making it difficult to employ any generalised approach, or tools or implementation platforms.

Taken together, all the foregoing factors argue strongly for the development of commercial nonlinear control packages in the same spirit as those available for (linear) model predictive control (MPC). Observe that, even though (i) linear MPC analysis and design techniques, obviously less complicated than nonlinear techniques, are still complicated enough compared to classical methods, and (ii) linear model development for MPC applications is still not a trivial task, commercial packages such as DMC and IDCOM (see Reference 20, Chapter 27, for a summary of other commercial MPC packages) have made the implementation of this technology much more widely accessible than would otherwise be possible.

Despite the obvious difficulties regarding 'standardisation' of model forms and design techniques, Continental Controls, Inc. has commercialised one nonlinear control package – MVC – with the claim that it could potentially do for nonlinear control what IDCOM and DMC did for linear model predictive control. One of the reported applications of this technology may be found in Reference 13. (See also Reference 14.)

In the next section we discuss the development and on-line performance of a nonlinear control system for an industrial process, to illustrate how the problems noted above – control technology, modelling and control system implementation – were addressed in this specific case.

10.3 Model predictive control of a spent acid recovery converter

10.3.1 The process

The process in question is the 'spent acid recovery' converter shown schematically in Figure 10.1. It consists of a series arrangement of four vanadium pentoxide fixed-bed reactors used to convert a cold feed of sulphur dioxide, (SO_2), oxygen, (O_2) and some inerts into SO_3. Because the reaction is highly exothermic, interstage

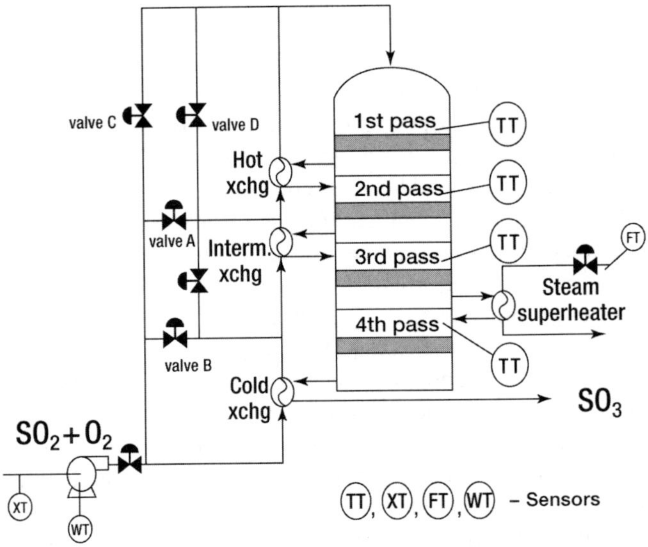

Figure 10.1 *Spent acid recovery converter*

cooling is provided primarily via heat exchange with the incoming cold feed, except between stages 3 and 4, where cooling is achieved via heat transfer to steam in a superheated steam generator.

10.3.2 Process operation objectives

Safe, reliable and economic process operation requires close regulation of the inlet temperatures of the first, second and third stages. In general, there is an 'optimum' inlet temperature for each stage (or pass) that will give rise to optimum conversion. These desired target values are determined by 'gas strength' (SO_2 concentration), production rate and the conversion achieved in the preceding passes. In addition, these temperatures must not fall below 410° (otherwise the reaction will be quenched) or rise above 600° (otherwise the catalyst active life will be shortened considerably).

Frequent fluctuations in feed conditions – the blower speed, gas strength (SO_2 concentration) and O_2 concentration – constitute the main obstacles to smooth process operation. Primarily to minimise yield losses, and to comply with strict environmental regulations on SO_2 emissions, these persistent disturbances must be rejected effectively and quickly. Ineffective process control has been responsible for low conversions, and low conversions result in both high SO_2 emission rates and high yield losses.

The indicated network of pipings, baffles and valves A, B and C provide the means for controlling the inlet temperatures through by-pass feeding of cold reactants. (For reasons that will soon become clear, only the valve openings – or 'valve loadings' – for valves A, B and C are available for manipulation; the valve loading of valve D is not.) For example, observe that increasing by-pass flow through valve C will reduce the first pass inlet temperature.

The dynamic characteristics induced by the network of valves can be quite complex. First, observe that the valves merely redistribute the feed, sending a portion directly as cold feed, and the rest through the various heat exchangers. A change in a single valve loading therefore affects not just the feed flow rate through that valve; it also affects the flow rate through all the other valves. These manipulated variables are therefore not entirely independent. Observe therefore that only three of the four valves can be manipulated independently. Next, consider, for the purpose of illustration, the effect of an increase in the valve C loading. The initial direct response will be a decrease in the first pass inlet temperature (as a result of increased cold feed bypass to this stage); but because the increased by-pass through valve C causes a concurrent decrease in the amount of cold feed distributed to the interstage heat exchangers, this action also results in an increase in the second and third pass inlet temperatures. This otherwise 'normal' process interaction is then complicated by secondary effects resulting from the fact that a reduction in the first pass inlet temperature ultimately causes a reduction in the exit temperature, which in turn causes a reduction in the inlet and outlet temperatures in the succeeding stages. The reduced temperature in all the stages then produces a tertiary effect in which the amount of the first stage feed preheating provided by the three interstage heat-exchangers is reduced, further reducing the first pass inlet temperature. This now starts another round of inlet temperature reductions with the potential for open-loop instability induced by the progressive cooling, and the possibility of quenching the reaction outright. Finally, as a result of the nonlinearity induced by the chemical reaction kinetics and the heat exchanger characteristics, a 'mirror image' decrease in the valve C loading will not give rise to a precise, 'mirror image' reverse net effect in inlet temperatures. To keep the process away from potentially unstable operating regimes, a lower constraint of 30 per cent is imposed on the valve loadings; the upper constraint of 100 per cent is physical.

The overall process objective may therefore be stated as follows:

> In the face of persistent process disturbances, control the inlet temperature for each of the first three passes to their respective prespecified desired target values, maintaining them between the operating constraints of 410°C, and 600°C at all times, with the loadings for valves A, B, and C constrained to lie between 30 and 100 per cent.

10.3.3 A control perspective of the process

The process variables may be categorised as follows:

- Output (controlled) variables:
    1. first pass inlet temperature

2. second pass inlet temperature
3. third pass inlet temperature.

- Input (manipulated) variables:
  1. valve A loading
  2. valve B loading
  3. valve C loading.

- Disturbance variables:
  1. SO_2 concentration
  2. O_2 concentration
  3. blower speed
  4. valve D loading.

As summarised above, the main control problems are caused by persistent disturbances, strong interactions among the process variables, constraints on both the input and output variables, and the process nonlinearities due to the reaction kinetics, heat transfer characteristics and the flow distribution network. The specific objective of the application is to develop an effective control system for this process, but the broader objective in this section is to use this specific application to illustrate various aspects of how nonlinear control can be applied on an industrial process.

10.3.4 Overall control strategy

The multivariable nature of the process, along with the process operating constraints, make this an ideal candidate for model predictive control (MPC); however, the severity of the process nonlinearities argues strongly for the application of nonlinear MPC instead of the more popular standard, linear version. The most important implications of this decision are as follows: technologically, this boils down – in principle – to obtaining a reasonable, nonlinear process model and a reliable nonlinear optimisation routine for performing the optimisation that lies at the heart of MPC. In practice, however, unlike with linear MPC, few theoretical results are available to guide the choice of critical design parameters such as the prediction horizon, the control move horizon and the various weights in the objective function. The nonlinear optimisation will thus have to be carried out with extra care. Also, unlike with linear MPC, no standard commercial packages were available at the time of this application (1991/92).

At the heart of the nonlinear model predictive control technique is the nonlinear process model, and based on the following three main points, the decision was made to obtain this model via input/output data correlation:

1. Not enough is known about certain critical details of the process to generate a first-principles model having sufficient integrity.

2. Even if the required fundamental process knowledge were available, the resulting first-principles model will be far too complicated for on-line optimisation-based control. Observe that, at the very least, such a model will consist of a combination of individual models for each subprocess making up the overall process: a gas distribution network model; a heat transfer model for the four heat exchangers; and a kinetic model for the four fixed-bed catalytic reactors. Each contributing model could conceivably consist of a system of several, coupled nonlinear partial differential equations, and the overall combination will clearly be far too complex for controller design.
3. From a process control perspective, the process is a 3×3 process with four disturbances; this process dimensionality is actually not so high as to render empirical modelling prohibitively time-consuming.

The issue of model structure selection in empirical nonlinear modelling is not trivial, and many factors influence each individual choice (see, for example, References 15 and 21). For this particular application, a recurrent neural network representation was chosen because of the flexibility of the neural network paradigm in general for representing arbitrary nonlinear input/output maps; the recurrent structure (as opposed to the standard feed-forward structure) was chosen in particular for improved long range prediction (see Reference 12), a critical requirement for model predictive control.

The overall control strategy is therefore to represent the process dynamics with a recurrent neural network, and to use this in a model predictive control framework in conjunction with a nonlinear optimiser. This control structure is shown in Figure 10.2.

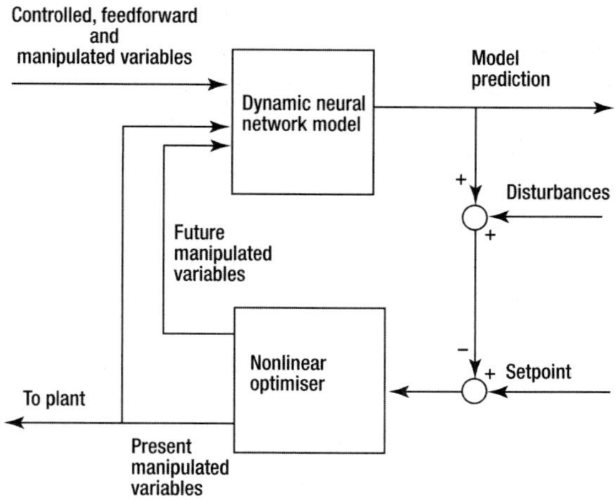

Figure 10.2 Control strategy

10.3.5 Process model development

A systematic procedure for nonlinear empirical model development involves the following steps [15]:

1. model structure selection
2. model identification (input sequence design; data collection and preconditioning; model parameter estimation)
3. model validation.

In this specific application, the selected model structure – a recurrent neural network – and the reasons for the choice have been presented. The next step – actual identification of the neural network model for the spent acid recovery converter – involves making decisions about the input sequences to be used for the model identification, implementing these input changes, collecting the sets of process response data, and analysing the collected input/output data sets.

The theoretical issues concerning input sequence design for nonlinear model identification remain largely unresolved (see, for example, Reference 15); much of what is done in practice is influenced mostly by sensible, but vague heuristics. For example, it is generally recommended that the magnitude of the inputs must be such that the desired region of operation is 'adequately covered' and that the 'frequency content' must be such that those aspects of the process that must be captured in the model are 'adequately excited'. Such heuristics and available theoretical results immediately rule out the typical inputs employed in industrial practice for linear model identification, i.e. single steps, single pulses and the PRBS; but there is as yet no comprehensive theory regarding 'optimum' input sequences for general nonlinear model identification.

In this specific case, therefore, the decision was to employ six-level, pseudo-random sequences (as opposed to the binary, i.e. two-level, sequences employed for linear systems) that span the 'normal' input range. From process operation data, and process knowledge, this 'normal' range was determined to be 30–80 per cent valve loadings. Because the 'dominant time constant' for the process is known to be approximately 40 min, the duration of each 'step change' in the sequence was fixed at 5 min, at the end of which the valve loading was switched to a different randomly drawn level. The total duration for each input sequence was fixed at 12 h.

Figure 10.3 shows the valve A loading input sequence and Figure 10.4 shows the observed responses in the first, second and third pass inlet temperatures, respectively. Similar responses were obtained from similar input changes in valves B and C.

Each process data set acquired during the plant tests was partitioned into two: one part for model development (the 'training set') and the other for model validation (the 'validation set'). The backpropagation-through-time algorithm was used to obtain the seven-input, three-output recurrent NN model from the plant data in the 'training set'. The final NN model architecture consisted of three layers and four nodes in the hidden layer, with unit time-delayed output feedback connections

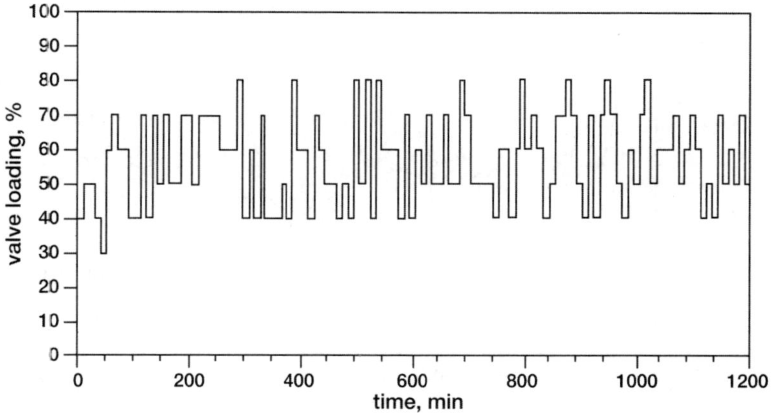

Figure 10.3 Identification input sequence for valve A loading

to the input layer. For additional details about the model development, see Reference 12. The performance of the resulting model is illustrated in Figure 10.5, where the long range, pure prediction of the first pass inlet temperature is compared with corresponding validation data. Comparable performance was observed from the other parts of the model.

10.3.6 Control system design and implementation

Conceptually, the nonlinear model predictive controller was implemented as shown in Figure 10.2: the NN model provided the long-range prediction, and 'ADS', a

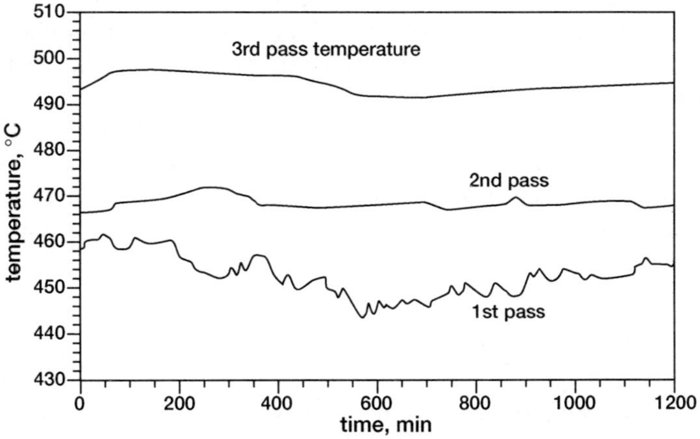

Figure 10.4 Temperature responses to changes in valve A loading

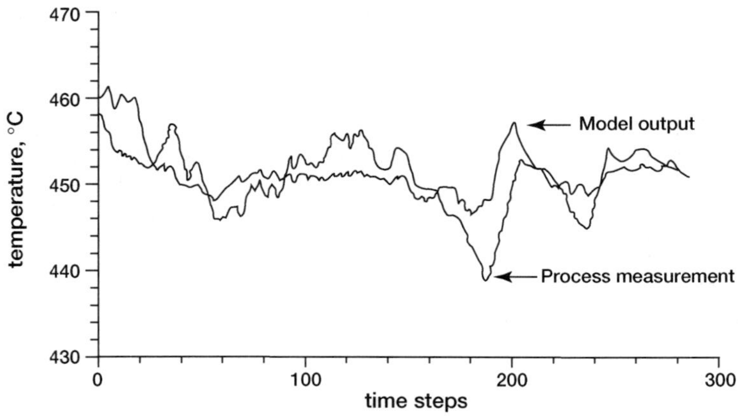

Figure 10.5 First pass inlet temperature prediction and validation data

public domain nonlinear optimisation routine (obtained from the Naval Postgraduate School in Monterey, CA) was used to determine optimal control action sequences. The model prediction and control sequence horizon lengths were chosen to be 20 and 5, respectively, with $\Delta t = 10\,\text{min}$. Additional details about the optimisation routine are available in Reference 12.

The actual implemetation of this nonlinear MPC scheme requires a few additional hardware and software considerations. Process operation data were collected and archived by a PDP 11/85 host computer interfaced to a dedicated DCS (distributed control system) through vendor-supplied software running on a MicroVAX system. The NN process model and the optimiser were deployed within an in-house expert system shell on the same MicroVAX computer. Apart from providing a convenient environment for integrating all the Fortran routines used to execute the modelling and the optimisation functions of the nonlinear MPC scheme, the expert system also performed two additional relatively simple, but critical, tasks: (i) it determined when it was time to execute the controller; and (ii) it checked the availability and validity of process data, and the 'reasonableness' of the computed control action.

At each control cycle, the desired setpoints computed for the valve loadings were sent from the expert system (in the microVAX) to the host computer; this was then communicated to the DCS, from where it was implemented on the actual process. The implementation hardware/software architecture is shown in Figure 10.6.

10.3.7 Control system performance

Figures 10.7–10.9 are representative of the actual closed-loop performance of the control system. Figure 10.7 shows the process output variables over a 24 h period during which the process was subject to the disturbances indicated in Figure 10.8.

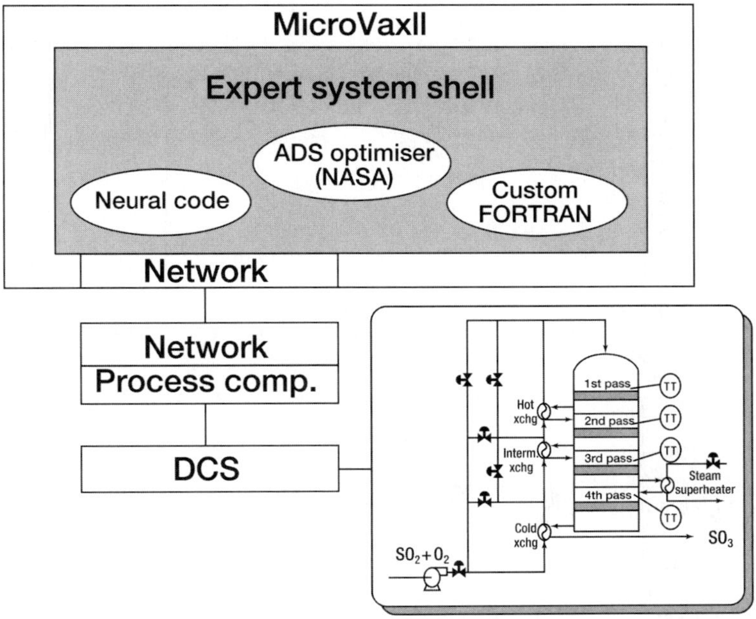

Figure 10.6 Control system implementation architecture

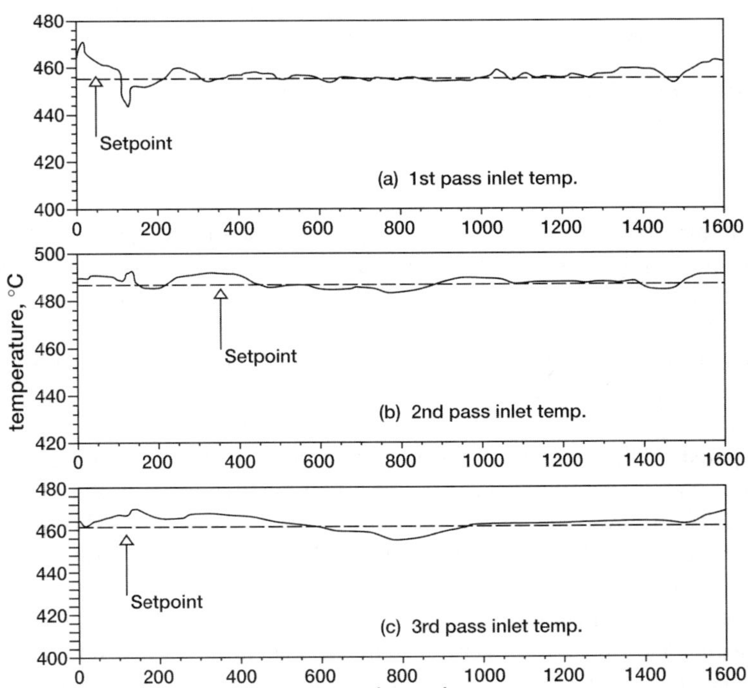

Figure 10.7 Closed-loop temperature responses

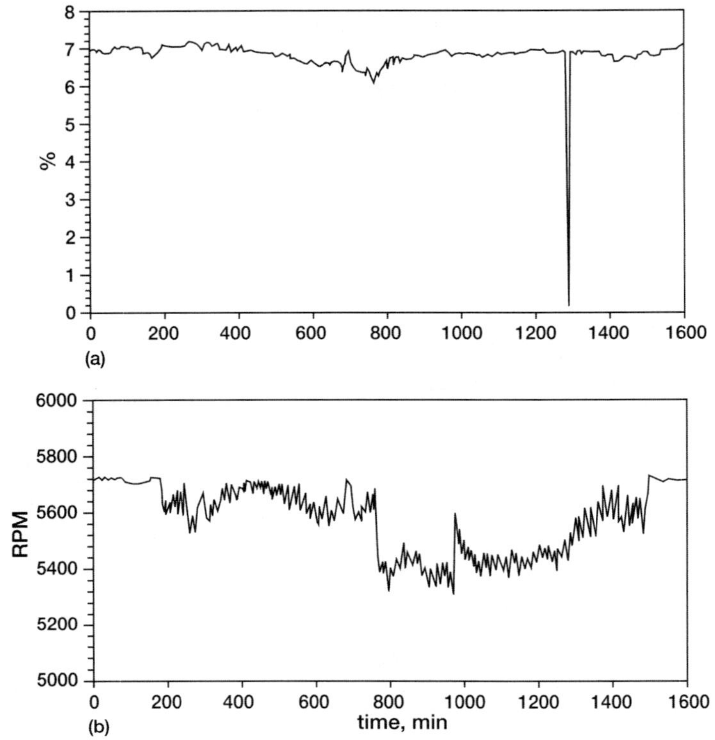

Figure 10.8 Process disturbances during closed-loop operation: (a) SO_2 concentration (b) blower speed

Between $t = 500$ and $t = 900$, the SO_2 concentration dropped by more than 15 per cent – by process operation standards, a significant feed disturbance; the indicated change in the blower speed (related to the process throughput) is also significant. In responding to these disturbances, the control scheme successfully maintained the inlet temperatures close to their respective desired setpoints, as shown in Figure 10.7, by implementing the control action sequences shown in Figure 10.9.

Compared with standard process operation prior to the implementation of this controller (not shown) the controller performed remarkably well. Observe that the 30–100 per cent constraint range was enforced for each of the valves during the entire period. The SO_2 concentration 'spike' that occurred at $t = 1300$ was due to the daily scheduled analyser calibration; observe, however, that such a clearly anomalous measurement did not affect the controller performance. This illustrates the effectiveness of the expert system in checking and validating process measurements before they are used in computing corrective control action. For additional details on the performance of the controller and a comparison to conventional control approaches, see Reference 12.

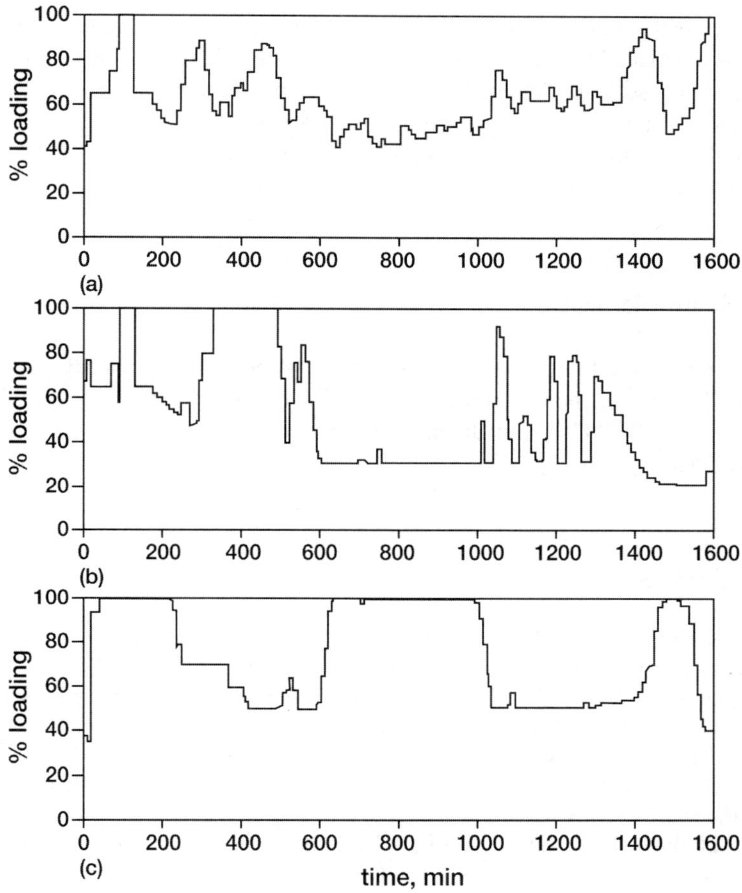

Figure 10.9 Implemented control actions: (a) valve A; (b) valve B; (c) valve C

10.4 Summary and conclusions

We have presented here one perspective of the 'many-sided' issues involved in the industrial application of nonlinear control, using the 'spent acid recovery' process as an illustrative case study of the successful design and implementation of one such industrial nonlinear control system.

Clearly, nonlinear control is becoming ever more relevant to industrial practice; the key issue now is essentially one of how best to identify and capture the stake presented by the ever-increasing demands on process operation. In this regard, by making the inevitable comparison with (linear) model predictive control and what has been primarily responsible for the significant impact it has had on industrial practice to date, it is not difficult to arrive at the following conclusion: the commercialisation of nonlinear control packages similar in spirit to those available for linear MPC will significantly increase the impact of nonlinear control on industrial practice. There are several obstacles to the widespread development and application of such packages; some of the most important have been noted. Nevertheless, that one such package is in fact already available is an encouraging sign that the potential exists for a significant increase in the application of nonlinear control techniques on many more actual industrial cases.

10.5 Acknowledgment

This chapter is based in part on an earlier paper jointly written with Ray Wright of The Dow Company, and presented at the 5th international conference on *Chemical process control* (CPC V) in January 1996. Ray's contributions are gratefully acknowledged.

10.6 References

1 KRAVARIS, C., and KANTOR, J.C.: 'Geometric methods for nonlinear process control', *Ind. Eng. Chem. Res.*, 1990, **29**, pp. 2295–2323
2 BEQUETTE, B.W.: 'Nonlinear control of chemical processes: a review', *Ind. Eng. Chem. Res.*, 1991, **30**, pp. 1391–1413
3 RAWLINGS, J.B., MEADOWS, E.S., and MUSKE, K.R.: 'Nonlinear model predictive control: a tutorial and survey'. Proceedings of ADCHEM'94, Kyoto, Japan, 1994
4 MEADOWS, E.S., and RAWLINGS, J.B.: 'Model predictive control,' in HENSON, M.A., and SEBORG, D.E. (Eds): 'Nonlinear process control' (Prentice-Hall, Englewood Cliffs, NJ, 1997)
5 SHINSKEY, F.G.: 'Process control systems' (McGraw-Hill, NY, 1979, 2nd edn)
6 WASSICK, J.M., and CAMP, D.T.: 'Internal model control of an industrial extruder'. Proceedings ACC, Atlanta, 1988, pp. 2347–52
7 LABOSSIERE, G.A., and LEE, P.L.: 'Model-based control of a blast furnace stove rig', *J. Process Control*, 1991, **1** (4), pp. 217–24
8 LEVINE, J., and ROUCHON, P.: 'Quality control of binary distillation columns via nonlinear aggregated models', *Automatica*, 1991, **27** (3), pp. 463–80
9 DORE, S.D., PERKINS, J.D., and KERSHENBAUM, L.S.: 'Application of geometric nonlinear control in the process industries: a case study'. *Control Engineering Practice*, 1995, **3** (3), pp. 397–402
10 WRIGHT, R.A., KRAVARIS, C., CAMP, D.T., and WASSICK, J.M.: 'Control of an industrial pH process using the strong acid equivalent'. Proceedings ACC, Chicago, 1992, pp. 620–29
11 WRIGHT, R.A., and KRAVARIS, C.: 'On-line identification and nonlinear control of an industrial pH process'. Proceedings ACC, Seattle, 1995, pp. 2657–61

12 TEMENG, K.O., SCHNELLE, P.D., and MCAVOY, T.J.: 'Model predictive control of an industrial packed bed reactor using neural networks', *J. Process Control*, 1995, **5** (1), pp. 19–27
13 BERKOWITZ, P.N., and GAMEZ, J.P.: 'Economic on-line optimization for liquids extraction and treating in gas processing plants'. Presented at the Gas Processors Association 74th Annual Convention, San Antonio, 1995
14 QIN, S.J., and BADGWELL, T.A.: 'An overview of nonlinear model predictive control applications', in ALLGÖWER, F., and ZHENG, A. (Eds): 'Nonlinear model predictive control' (Birkhauser, Switzerland, 2000), pp. 369–92
15 PEARSON, R.K., and OGUNNAIKE, B.A.: 'Nonlinear process identification,' in HENSON, M.A., and SEBORG, D.E. (Eds): 'Nonlinear process control' (Prentice-Hall, Englewood Cliffs, NJ, 1997), chapter 2, pp. 11–110
16 PSICHOGIOS, D.C., and UNGAR, L.H.: 'A hybrid neural network – first principles approach to process modelling', *A.I.Ch.E.Journal*, 1992, **38**, p. 1499
17 TULLEKEN, H.J.A.F.: 'Grey-box modelling and identification using physical knowledge and Bayesian techniques', *Automatica*, 1993, **29**, pp. 285–308
18 LINDSKOG, P., and LJUNG, L.: 'Tools for semi-physical modelling'. Preprints IFAC Symposium on Systems Identification, 1994, vol. 3, pp. 237–42
19 OGUNNAIKE, B.A.: 'Application of hybrid modelling in control system analysis and design for an industrial low-boiler column'. Proceedings European Control Conference, Rome, 1995, pp. 2239–344
20 OGUNNAIKE, B.A., and RAY, W.H.: 'Process dynamics, modelling, and control' (Oxford University Press, NY, 1994)
21 PEARSON, R.K.: 'Nonlinear input/output modelling', *J. Process. Control*, 1995, **5** (4), pp. 197–211

Chapter 11

Nonlinear model based predictive control using multiple local models

Shane Townsend and George W. Irwin

Abstract

The chapter describes nonlinear generalised predictive control (GPC) where the internal linear plant model is replaced by a local model (LM) network representation.

While artificial neural networks can model highly complex, nonlinear dynamical systems, they produce black box models. This has led to significant interest in LM networks to represent a nonlinear dynamical process by a set of locally valid and simpler submodels. The LM network structure, its interpretation and training aspects are introduced. The network was constructed from local autoregressive with external input (ARX) models and trained using hybrid learning.

Two alternative methods of exploiting the LM network within a generalised predictive control (GPC) framework for long-range, nonlinear model predictive control are described. The first consists of a network of predictive controllers, each designed around one of the local models. The output of each controller is passed through a validity function and summed to form the input to the plant. The second approach uses a single predictive controller, which extracts a local model from the LM network to represent the process at each controller sample instant.

Simulation studies for a pH neutralisation process show the excellent nonlinear modelling characteristics of the LM network. Both nonlinear model predictive controllers gave excellent tracking and disturbance rejection results and improved performance compared with conventional linear GPC.

11.1 Introduction

Traditionally, process control involves the use of linear control techniques. Although linearity may not be a good approximation to the actual process behaviour, it has been proved successful when the plant remains close to an operating point or when the nonlinearity is relatively minor. In practice, the proportional-integral-derivative (PID) controller is the most widespread, being deployed in most industrial control loops. This algorithm is well understood, relatively easy to tune and has some extensions to handle nonlinearities (e.g. anti-windup) [1–3].

Advanced strategies used for process control [4] include adaptive control, statistical process control, internal model control and, of course, model based predictive control (MPC). Since the early 1980s [5, 6] a number of alternative MPC strategies have emerged including generalised predictive control (GPC) [7, 8] (which incorporates a linear controlled autoregressive with integrated moving average (CARIMA) plant model) and model algorithmic control [9]; (which utilises an impulse response model in the control law). If the process is linear, stable with no constraints and the desired process output is constant for the foreseeable horizon, then all of these controllers generally yield approximately the same result. Thus, all of these linear control laws have the same structure with a sufficient number of degrees of freedom (after some manipulation) [10]. MPC has enjoyed considerable commercial success. The results of a 1995 survey [11] reported over 2200 applications worldwide, mainly for large oil refineries and petrochemical plants, with rather less impact in other process industries.

Since the MPC control law relies on a *linear* model of the process generated at a particular operating point, the controller's internal model will be less representative of the dynamics of a nonlinear process as it moves away from this point. This in turn reduces the robustness of the closed-loop system, hence the significance of the nonlinear MPC control strategies described in the present book. The field of nonlinear MPC is now well established and in a 1998 survey [12] of five vendors this technology produced a total of 86 industrial applications.

Using a neural network to learn the plant model from operational process data for nonlinear MPC is one solution. A number of alternative architectures have been studied such as back-propagation networks [13] and dynamic neural networks [14]. Applications reported in the research literature include a packed-bed reactor [15], a distillation column [16] and in-line neutralisation [17]. An alternative is to use a set of local models to accommodate local operating regimes [18–20]. This is attractive since the plant model used for control provides a transparent plant representation as compared to 'back-box' neural networks.

The present chapter describes how this recent nonlinear modelling technique, which retains some of the insights obtained from linear systems, can be integrated within an MPC framework. The nonlinear model used is called a local model (LM) network [21] and is built up from a set of locally valid submodels. A global plant representation is then formed using multiple models spread throughout the operating space of the nonlinear process.

Two methods are proposed for nonlinear MPC based on an identified LM network model of the plant. The first is based on a network of controllers, each designed and tuned about an individual locally valid submodel. The controller outputs are each passed through a validation function with a magnitude dependent upon the current operating point, before being summed to form the plant input. In the second approach, a GPC controller is supplied with local models extracted from the LM network at different operating points, thereby incorporating the nonlinear model within the controller. This technique avoids the use of nonlinear optimisation normally associated with MBPC methods that involve the use of full nonlinear neural network models [16]. Here the LM network forms a global plant model from a set of locally valid submodels [22]. The model extracted by the controller at each sample instant is locally representative of the process at that operating point.

Simulation results for a pH neutralisation process show the excellent nonlinear modelling properties of the LM network. Comparative control studies produce good tracking and disturbance rejection results for both nonlinear MPC schemes. Further, the results suggest an improvement over conventional linear GPC.

11.2 Local model networks

All linear models or controllers will have a limited operating range within which they are accurate or perform adequately well. The validity of linearisation, modelling assumptions, stability properties and experimental conditions all affect the effective operating range of the model/controller in practice. A local model/controller is one where its useful operating range is less than that of the full range of operating conditions, as opposed to a global one, which operates over all of the expected conditions. In general, a global model/controller is required and LM networks provide a useful approach to achieve this goal.

In contrast to neural networks such as the Multilayer Perceptron (MLP) or Radial Basis Function (RBF), the local model (LM) network forms a global plant model from a set of locally valid submodels [22]. The general feedforward structure of the LM network contains submodels that could therefore be neural networks or even simple linear plant models. Thus, linear models and any *a priori* information from a physical modelling exercise can easily be incorporated within this structure. The outputs of each submodel are passed through a local processing function that effectively acts to generate a window of validity for the model in question. These nonlinear weighting functions utilise only a subset of the available modelling data to generate the desired partitioning of the model space. The resultant localised outputs are then combined as a weighted sum at the model output node. Figure 11.1 shows a diagram of an LM network structure, the variables of which are explained below.

The essence of the operating regime based approach is to decompose the operating space into regimes, where models (of relative simplicity compared to the

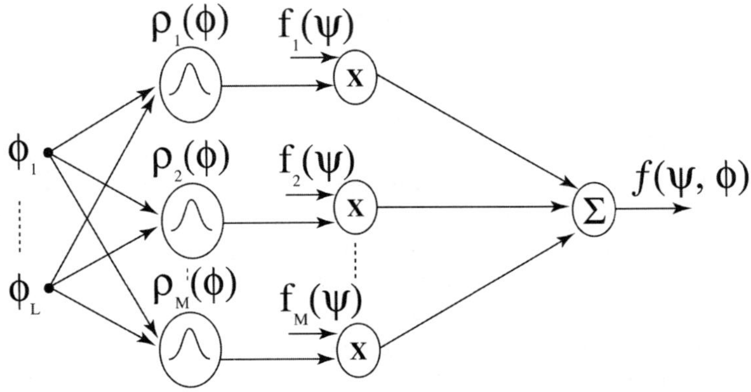

Figure 11.1 Local model (LM) network

original nonlinear process) are specified which are adequate approximations to the dynamical relationship within any particular regime. A compromise exists between the number and size of operating regimes and the complexity of the corresponding local model. Decomposition into a few large operating regimes will require more complex local models than decomposition into numerous small operating ones [21]. Provided that the functional relationship to be modelled is smooth, approximations based on operating regimes and local models can be made arbitrarily accurate either by making the decomposition sufficiently fine, or by making the local models sufficiently complex [23]. This type of function approximation brings with it the curse-of-dimensionality: 'With an increasing number of variables on which the function depends, the number of partitions required in a uniform partitioning will increase exponentially' [21, pp. 3–72]. Accordingly, a uniform partitioning of the operating space is undesirable and unrealistic for anything other than low complexity problems [21, 24].

The LM network can be interpreted as a generalised form of an RBF neural network (see Figure 11.1) in the sense that the basis functions now multiply general functions of the inputs. The network output is given by the following equation:

$$\hat{y}(k+1) = \hat{F}(\underline{\psi}(k), \underline{\phi}(k)) = \sum_{i=1}^{M} f_i(\underline{\psi}(k)) \cdot \rho_i(\underline{\phi}(k)) \tag{1}$$

Here the M local models $f_i(\underline{\psi})$ are linear or nonlinear functions of the measurement vector $\underline{\psi}$, and are multiplied by a basis function $\rho_i(\underline{\phi})$ that is an function of the current operating region vector, $\underline{\phi}$. The latter does not necessarily need to be the full model input vector, $\underline{\psi}$, but can be a subset of the measurement data available. For comparison note that in the RBF neural network the functions $f_i(\underline{\psi})$ are constants and the basis functions $\rho_i(\underline{\phi})$ are radial. The basis functions $\rho_i(\underline{\phi})$ in (1) are commonly chosen to be normalised Gaussian functions, c_i and σ_i being the centres and widths, respectively, such that

$$\rho_i(\underline{\phi}) = \frac{\exp(-\|\underline{\phi} - c_i\|^2/2\sigma_i^2}{\sum_{i=1}^{M} \exp(-\|\underline{\phi} - c_i\|^2/2\sigma_i^2)} \qquad i = 1, 2, \ldots, M \tag{2}$$

This function gives a value close to 1 in parts of $\underline{\phi}$ where the local f_i is a good approximation to the unknown F and a value close to 0 elsewhere. If, for example, the local models are of the affine ARX form,

$$\begin{aligned} f_i(\underline{\psi}) = {} & b_{oi}u(k) + b_{1i}u(k-1) + \cdots + b_{si}u(k-s) + \cdots + a_{1i}y(k) \\ & + a_{2i}y(k-1) + \cdots + a_{(r+1)i}y(k-r) + c_i \end{aligned} \tag{3}$$

where r, s are the orders of $y(k)$ and $u(k)$, respectively, then the LM network constitutes a nonlinear ARX plant model as follows:

$$\begin{aligned} & \hat{y}(k+1) = \hat{F}(\underline{\psi}(k), \underline{\phi}(k)); \\ & \quad \underline{\psi}(k) = [y(k), y(k-1), \ldots, y(k-r), \ldots, u(k), u(k-1), \ldots, \\ & \quad u(k-s), 1]^T; \ \underline{\phi}(k) \subset \underline{\psi}(k) \end{aligned} \tag{4}$$

Combining (1) and (4) yields the following expansion:

$$\begin{aligned} \hat{y}(k+1) = {} & B_0u(k) + B_1u(k-1) + \cdots + B_su(k-s) + \cdots + A_1y(k) \\ & + A_2y(k-1) + \cdots + A_{r+1}y(k-r) + C \end{aligned} \tag{5}$$

The resulting model is still ARX in structure with the parameters A_i and B_i defined at each operating point by

$$B_i = \sum_{j=1}^{M} \rho_j(\underline{\phi}(k))b_{ij} \quad A_i = \sum_{j=1}^{M} \rho_j(\underline{\phi}(k))a_{ij} \quad C = \sum_{j=1}^{M} \rho_j(\underline{\phi}(k))c_j \tag{6}$$

The underlying assumption in the local modelling strategy is that the plant to be modelled undergoes significant changes in operating conditions. For most batch and continuous processes in the chemical, biotechnological and power industries, definite regimes can be identified during procedures such as start-up, shutdown and product shifts. Incorporating simpler models in each operating region can reduce the complexity of the overall model. For example, local state-space and ARMAX models can be formed using localised perturbation signals and then interpolating to produce global nonlinear state-space and NARMAX (nonlinear ARMAX) representations.

The identification of local operating regimes for an unknown plant can be difficult in general. Any such identification strategy has to take into account the complexity of the target mapping, the representational ability of the local models associated with the basis functions and the availability of the data. The problem is, therefore, to identify those variables which describe the system operating behaviour. *A priori* knowledge of the plant can be used at this stage. When little knowledge of the actual regimes exists, it may be beneficial to use unsupervised learning methods, such as *k*-means clustering and nearest neighbours, to give an initial estimate of the normalised Gaussian interpolation regions. Such clustering methods are valid in this case since many plants tend to operate in distinct regions.

A common approach to regime decomposition is to divide the operating space using a uniform multidimensional grid, which may be retained as the final decomposition or employed as an initialisation for supervised training techniques. However, this generally results in a large number of regimes due to the curse of dimensionality. Hierarchical/constructive techniques attempt to form a more parsimonious decomposition of the operating space employing a top-down approach, which gives rise to a tree-like partitioning of the operating space. A prime example of this is the *k-d* tree decomposition, which applies a series of splits across a subspace to be partitioned with the number of regimes being increased by one with each split. Johansen and Foss [25, 26] and Nelles and co-workers [19, 27] have proposed similar construction methods. Each regime is defined with a Gaussian basis function placed at its centre with the standard deviation made proportional to the width of the regime along each dimension. The operating regime is split slightly differently in each case. The Nelles algorithm forms a new split across the regime where the current fit to the data is the poorest.

Alternatively Johansen and Foss choose the split which results in the best new fit to the data. Their algorithm must consequently evaluate all the new splits, thereby giving a better chance of obtaining the best new fit to the data. By contrast the Nelles strategy only evaluates splits over a single regime and is much less computationally intensive [20].

Successful applications of LM networks for nonlinear identification have been reported in the biotechnology and chemical engineering industries [18, 28] and more recently for a turbogenerator [29].

As with conventional neural networks, training is a crucial issue since for LM networks there is the added complexity of identifying the local models as well as the parameters of the interpolation functions. This chapter employs a hybrid learning approach for LM networks built from ARX local models and normalised Gaussian basis functions. Singular value decomposition (SVD) is used to identify the local linear models in conjunction with quasi-Newton optimisation for determining the centres and widths of the interpolation functions. To avoid over-training problems, which arise in nonlinear dynamic modelling with noisy data, the SVD minimises a one-step-ahead prediction error while the nonlinear optimisation is performed on a parallel model based error [30].

11.3 Nonlinear model based predictive control

The approach adopted here to nonlinear MPC is to incorporate the LM network into a model predictive control structure. The aim is to overcome some of the drawbacks of the more usual approach based on a complete nonlinear plant model. For example, the requirement for nonlinear optimisation to minimise the cost function for control, which inevitably suffers from a large computational penalty, may be avoided. In particular, this section describes how the LM network representation of a plant can be integrated into a generalised predictive control (GPC) framework as a basis for long-range, nonlinear predictive control.

GPC is an internal model control strategy based on a linearised plant and has been used successfully for industrial process control for some time. A discrete-time model of the process is employed to predict the behaviour of the controlled outputs over a finite time-horizon at each sample instant. This predicted process behaviour is then used to find a finite sequence of control moves that minimises a particular quadratic objective function without violating prespecified constraints. Only the first element of this control sequence is presented to the plant and at the next sample instant the procedure is repeated using the latest measured process information within a receding horizon. Full details for the basic algorithm may be found in Clarke *et al.* [7].

Two different nonlinear control methods are now described which take advantage of the transparency and structure of the LM network in conjunction with a GPC control strategy.

11.3.1 Local controller generalised predictive control (LC-GPC)

This technique is very similar to gain scheduling, which is probably the most commonly used approach to the control of highly nonlinear systems [31]. A gain scheduling controller is constructed by interpolating between the members of a family of linear controllers. Simple design, tuning and relatively low computational burden means that this remains a very favourable control strategy amongst practising control engineers.

The control structure here consists of the family of controllers and the scheduler. At each sample instant the latter decides which controller, or combination of controllers, to apply to the process. Generally, the controllers are tuned about a model obtained from experiments about a particular equilibrium point, since linear models and controllers are quite well understood.

This creates a further dilemma about the placement of these equilibrium points and associated controllers across the operating regime in such a way as to cover the operating range of the process but without using so many controllers that the design becomes excessively complex. It has been suggested that the realisation of a gain-scheduled controller should be chosen to satisfy a local linear equivalence condition, that is, the linearisation of the gain scheduled controller, at an equilibrium operating point, should correspond to the associated member of the family of linear controllers which was used to realise the controller. However, the dynamic behaviour of the resulting nonlinear controller can be strongly dependent on the realisation adopted and, since the underlying nonlinear process may exhibit quite different dynamic behaviour, it has been suggested that this condition is inadequate [32]. The scheduling algorithm plus the process variables used should clearly be selected so as to capture the process nonlinearity.

As already described, the LM network consists of a set of locally valid submodels together with an appropriate interpolation function. A controller is now designed about each of the local models. The output of each controller is then passed through the interpolation function which effectively generates a window of validity for each of the individual controllers. The interpolated outputs are then summed and used to supply the control commands to the process. The resultant LC-GPC structure is as shown in Figure 11.2 for the pH application to be described in the next section. The interpolation function effectively smoothes the transition between each of the local controllers. In addition, the transparency of the nonlinear control algorithm is improved as the operating space is covered using controllers rather than models. Since the LM network has been optimised for the number of models used, it is known in advance that the realisation has a minimum number of controllers as well as being based on a good global representation of the underlying nonlinear process.

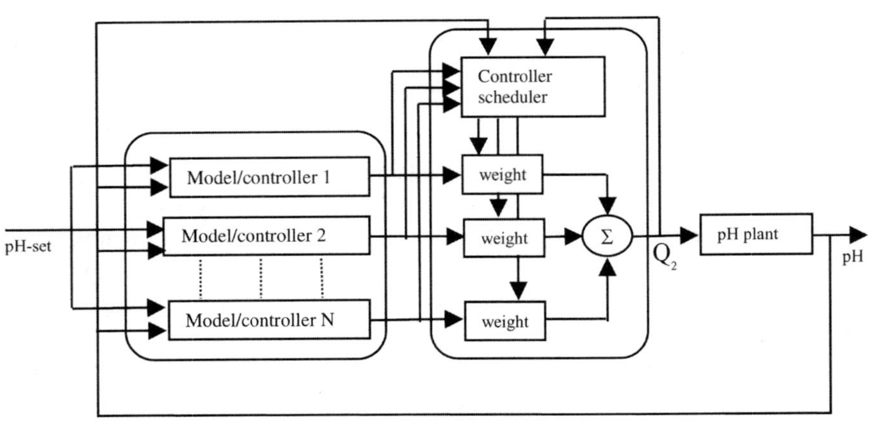

Figure 11.2 Local controller generalised predictive control scheme

11.3.2 Local model generalised predictive control (LM-GPC)

As described earlier, the LM network provides a global nonlinear plant representation from a set of locally valid ARX models together with an interpolation function. The latter generates activation weights for each model, which have a value close to one in parts of the operating space where the local model is a good approximation and a value approaching zero elsewhere.

In Section 11.3.1 the controller structure/network exploited the operating regime decomposition by placing a controller at each local model. An alternative way of developing another nonlinear controller is to use the same operating regime based

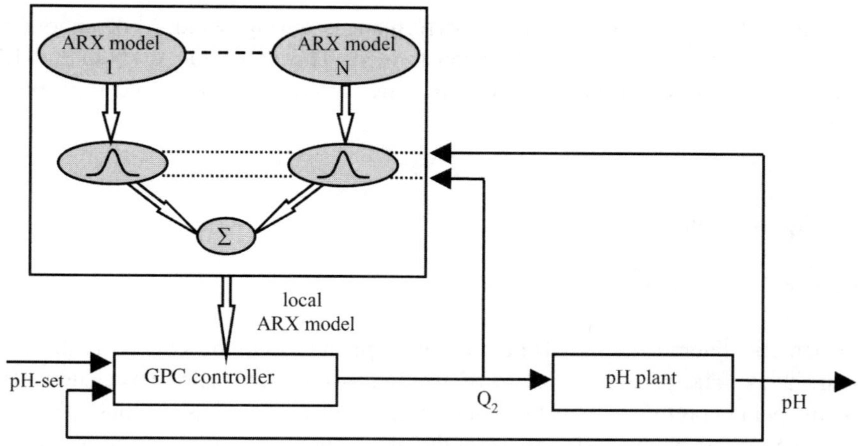

Figure 11.3 Local model generalised predictive control scheme

model directly with a model based control framework such as GPC. The resultant global controller is expected to give better plant-wide control performance than the equivalent linear controller, simply because global modelling information may be used to determine the control input at each sample time. The closed-loop performance, stability and robustness are then all directly related to both the quality of the identified model and the general properties of GPC control [21].

The parameters of each local model in the LM network are passed through the interpolation function to produce an overall nonlinear ARX model of the plant. At any particular operating point this model may be assumed to be a locally valid representation of the plant. However, the affine (as opposed to linear) form in (3) is unavoidable to achieve accurate modelling [33] if each local model is of a positional nature. Currently research is being carried out to form an LM network based on velocity models which are then not required to be affine locally [34]. At each operating point this affine term is constant and is therefore regarded as a constant disturbance to the process which is allowed for in the GPC control algorithm. This allows the affine term to be neglected, and the conversion of the ARX model to incremental form is now quite straightforward. It is assumed to constitute a linear representation of the process at that time instant and may then be used by a GPC controller to represent the process dynamics locally. Since the model is regarded as linear and valid, the control sequence can be solved analytically at each sample instant, assuming that no constraints are to be applied. The resultant LM-GPC structure for the pH application to be described in Section 4 is illustrated in Figure 11.3. The operating vector in this case was taken to be the output value of the process and the controller input to the process.

The controllers described in Sections 11.3.1 and 11.3.2 were applied to a simulation of a pH neutralisation pilot plant to be outlined in Section 11.4.1. Section 11.4.2 describes how an LM network was developed for this plant, and

Section 11.4.3 discusses the control performances during testing. For simplicity, both controllers were SISO and unconstrained. The extension of both the LM network and the controllers for multivariate systems is the subject of future work.

11.4 Application

11.4.1 pH neutralisation pilot plant

A schematic diagram for the pH neutralisation plant used in this chapter is shown in Figure 11.4. The process consists of weak concentration acid, base and buffer streams being continuously mixed within a reaction vessel whose effluent pH value is measured. The objective is to control the pH value of the outlet stream by varying the inlet base flow rate, Q_2. The acid and buffer flow rates of the process, Q_1 and Q_3, respectively, are controlled using peristaltic pumps. The outlet flow rate is dependent upon the fluid height within the vessel and the position of the outlet valve, which is set manually. The nominal parameters for this system are summarised in Table 11.1. The simulation used in this work is described in greater detail in Reference 35.

11.4.2 Identification

For this particular plant, the static process gain between base flow rate and pH (i.e. the slope of the titration curve – see Figure 11.5) varies considerably as the pH and base flow change. It can be seen from the titration curve that there are five regions in which the gain is nearly constant. Also, within each region linear identification experiments suggested that a second-order dynamic model was sufficient to describe the behaviour of the process. An LM network with five local second-order, linear ARX models was therefore constructed to give:

$$\begin{aligned} p\hat{H}(k+1) &= \sum_{i=1}^{5} f_i(\underline{\psi})\rho_i(\underline{\phi}) \\ \underline{\psi}(k) &= [pH(k), pH(k-1), q_2(k-d), q_2(k-1-d)]^T \\ \underline{\phi}(k) &= [pH(k), q_2(k-1-d)]^T \end{aligned} \tag{7}$$

In (7), pH(k)and $q_2(k)$ are the process output (pH) and the process input (base flow rate), respectively, at time k, while d is the system delay. Both pH and base flow rate were used to specify the operating point. Training and test data were generated using perturbations on the inlet base flow rate. The LM network was subsequently trained using a hybrid optimisation method [30]. As already described in Section 11.2, SVD was used to identify the local linear models in conjunction with quasi-Newton optimisation for determining the centres and widths of the interpolation

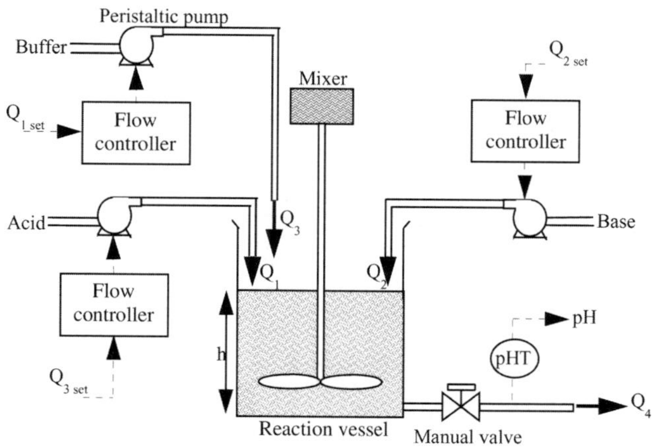

Figure 11.4 The pH neutralisation pilot plant

functions. To avoid over-training problems, which arise in nonlinear dynamic modelling with noisy data, the SVD minimises a one-step-ahead prediction error (series-parallel model), while the nonlinear optimisation is performed on a parallel model based error (see Figure 11.6). The generalisation performance of the LM network is made up of two conflicting sources of error, which is known as the bias/variance problem [36]. The bias term arises from the model structure being too simple to capture the overall function dynamics, whereas the variance term arises when the model structure is too flexible such that the model learns the noise on the data. If the latter occurs, high frequency components not present in the underlying function are introduced and the network is said to be over-fitted/over-trained, leading to a badly conditioned network. In general, to prevent over-fitting, regularisation is applied to the training to improve the conditioning of the model by introducing an additional term in the optimisation cost function [30]. Figure 11.7 shows that excellent nonlinear predictive models were obtained for all training/test sets.

Figure 11.8 (a–e) shows the interpolation functions for the associated LM network. These 3-D graphs give a visual representation of how the operating space is spilt up between the different local models. The *x*- and *y*-axes contain

Table 11.1 Nominal operating conditions of the pilot pH neutralisation plant

Acid stream 0.003M NHO_3	$Q_1 = 16.6\,\mathrm{ml/s}$
Buffer stream 0.003M $NaHCO_3$	$Q_2 = 15.6\,\mathrm{ml/s}$
Base stream 0.003M NaOH	$Q_3 = 0.55\,\mathrm{ml/s}$
0.00005M $NaHCO_3$	

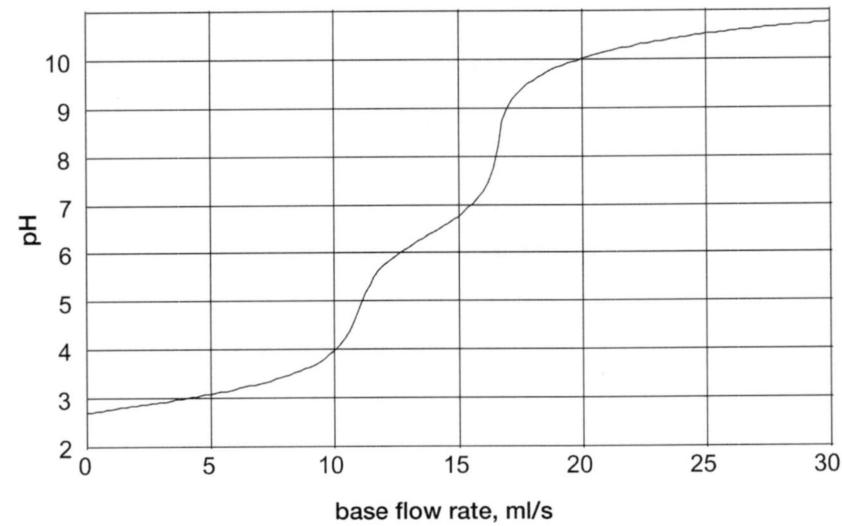

Figure 11.5 Titration curve

the scheduling variables (outlet stream pH value and base flow rate), and the associated weight or height for each operating point lies on the *z*-axis. It can be seen that the operating space is split up between the five models in such a way so as not to violate partition of unity. Models 1, 3 and 5 tend to cover most of the operating space, with models 2 and 4 added to improve the model fit in these areas. As already mentioned, the interpretation of this network structure is much more transparent than that of other nonlinear models and, in particular, neural networks.

11.4.3 Control

LC-GPC: Here the nonlinear controller consists of a set of local controllers, each one of which is designed about one of the local models. At each sample instant the outlet stream pH value and the input stream base flow rate were fed to the interpolation function of the LM network, which in turn generates the activation weights for each of the local controllers. Each of the five local controller outputs was multiplied by these activation weights before summing to form the process input as illustrated in Figure 11.2. Each local controller was assumed to be linear and hence the control sequence for each could be solved analytically. Note, however, that the interpolation function is nonlinear and therefore the summation of the interpolated outputs is also nonlinear.

LM-GPC: In this case the LM network for the pH process was used with a GPC algorithm for control purposes. At each sample instant the outlet stream pH value and the input stream base flow rate (Q_2) were fed to the interpolation function of

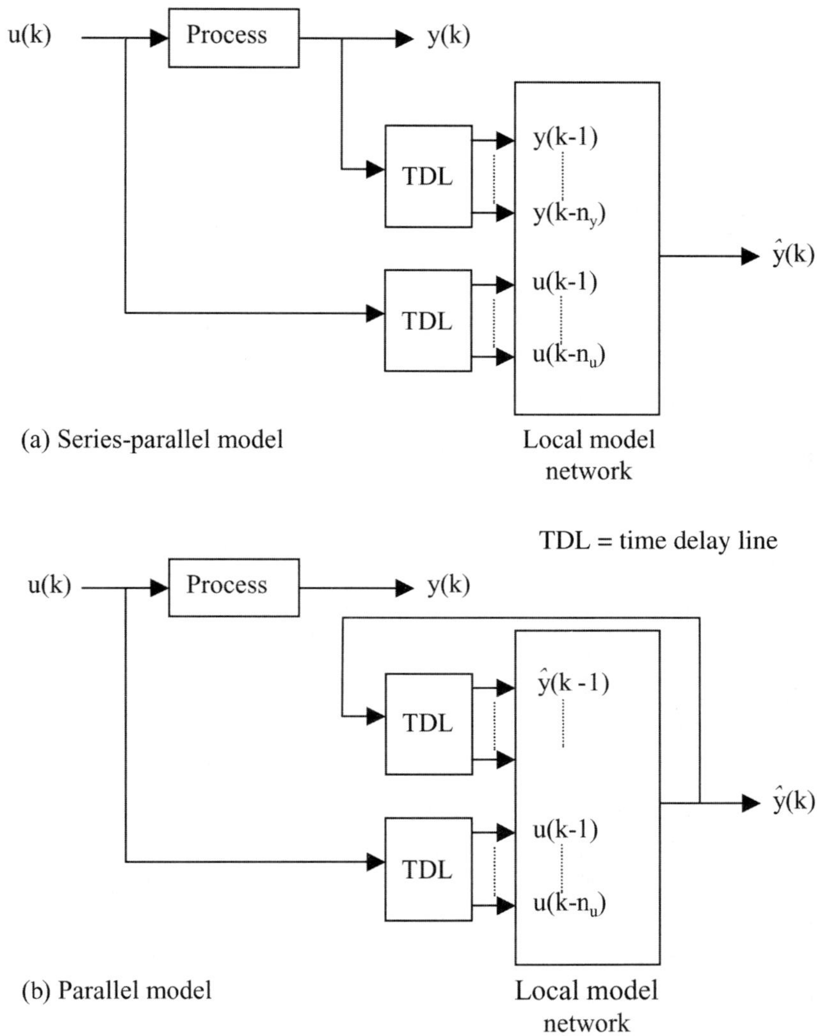

Figure 11.6 Series-parallel and parallel model formulations

the LM network. Each of the five sets of local ARX model parameters was then passed through this interpolation function to form a local ARX model, which accurately represented the process around that particular operating point. This local model may be assumed linear and is used by the GPC controller as shown in Figure 11.3.

Interestingly, a somewhat similar method to this has also been implemented upon a temperature plant [37]. This uses a fuzzy identification algorithm to adaptively update the internal model used by the dynamic matrix. Likewise, Fischer *et al.* [38] extract step response models from Takagi-Sugeno fuzzy systems to implement a

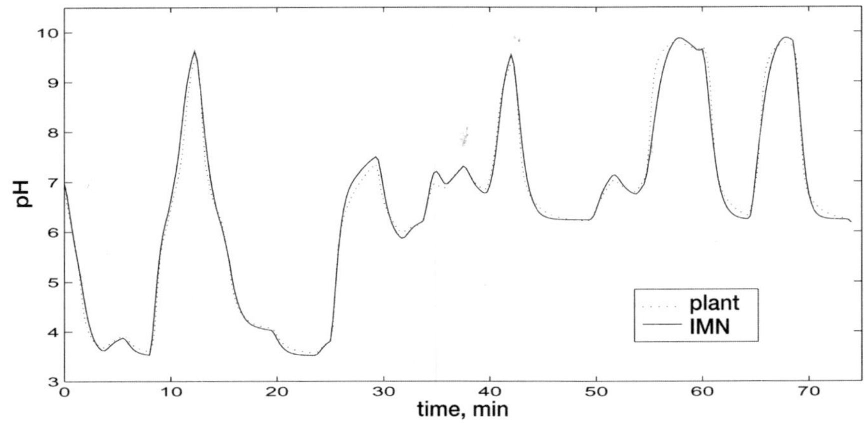

(a)

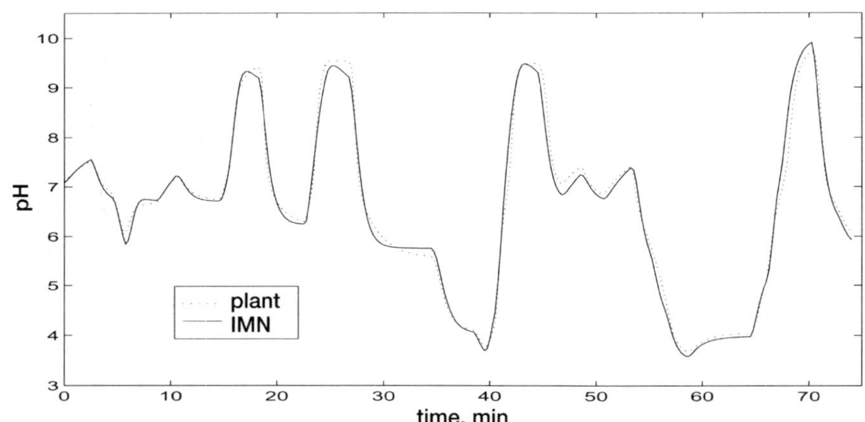

(b)

Figure 11.7 pH/base scheduled local model network parallel model output for (a) training set and (b) test set

nonlinear predictive controller. However, in this case the process step response is updated only when there is a setpoint change.

The two new nonlinear control strategies were tested for tracking and disturbance performance and compared to linear GPC.

Setpoint tracking: For this test the process was set up under steady-state conditions, with an outlet pH value of 4.0. At T equal to 300 s the setpoint was incremented and thereafter every 1000 s until it reached 10.0. This was done in order to move the pH process across a wide operating space in which the process

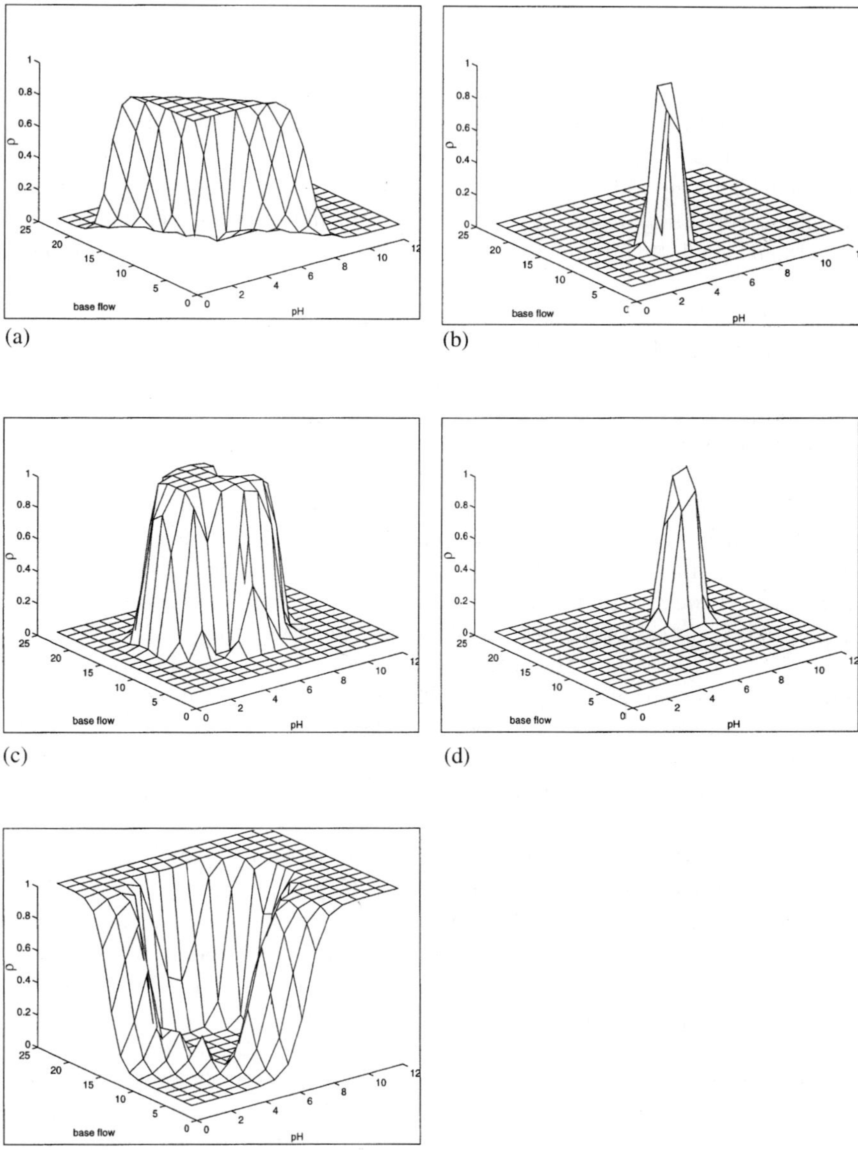

Figure 11.8 *pH/base scheduled interpolation functions for: (a) model 1; (b) model 2; (c) model 3; (d) model 4; (e) model 5*

gain varies significantly. The results of this test are shown in Figure 11.9. It is readily apparent that the linear GPC controller has great difficulty in controlling the outlet pH value at 8.0. This is because its internal model was generated at a pH value equal to 7.0 where the plant gain is moderate while the operating region around pH equal to 8.0 corresponds to a high gain region. The nonlinear GPC controllers show no difficulty in tracking the setpoint, although there is some indication of a slower response time for some of the step changes. Overall there seems to be very little difference between the two nonlinear controllers during this test.

The LM-GPC strategy requires only one set of tuning parameters. The internal model of a single GPC controller is updated at each sample instant. It is therefore necessary to tune this nonlinear controller so that the highest frequency dynamics of the process always take precedence. This is seen to be a limiting factor as when the pH process moves to an operating point with a lower gain, the tuning parameters are such that the best control performance may not be achieved.

For this application, the LC-GPC controller consisted of five local GPC controllers, each with a set of tuning parameters. Since the controllers may be tuned independently, this avoids the effect of high frequency dynamics taking precedence. However, tuning these controllers was not easy due to both their number and the effect of the normalised basis functions of the LMN [39].

Disturbance rejection: The pH neutralisation process is highly nonlinear, containing a wide variation in process gain throughout its operating space. For this reason, it was decided to test the controllers at two distinct operating points (pH equal to 7.0 and 9.0) with different process gains. Disturbances were introduced to the system by reducing the buffer flow rate from its nominal flow rate of 0.55 ml/s while the process was in steady state.

In Figure 11.10 the minimum buffer flow rate is shown, which represents the maximum disturbance that may be tolerated by the process before large oscillations occur, for all three controllers at both operating points. This is necessary, as all the controllers should give a similar performance when the setpoint is 7.0, since all three controllers are tuned for similar dynamics. But the nonlinear GPC controllers should exhibit superior disturbance rejection when the setpoint is 9.0. The tuning parameters of the linear controller are now no longer valid when the setpoint is 9.0 as they were specified at a setpoint of 7.0. Figure 11.10 verifies that this is the case. The controllers show similar performance when the setpoint is 7.0, with the nonlinear controllers being slightly better. However, when the setpoint was moved to 9.0, the nonlinear controllers significantly outperformed the linear GPC.

11.5 Discussion and conclusions

GPC was originally conceived as a general-purpose algorithm for the stable control of the majority of real processes. It can be used with nonminimum-phase plant, open-loop unstable plant, plant with variable or unknown dead time and plant with unknown order.

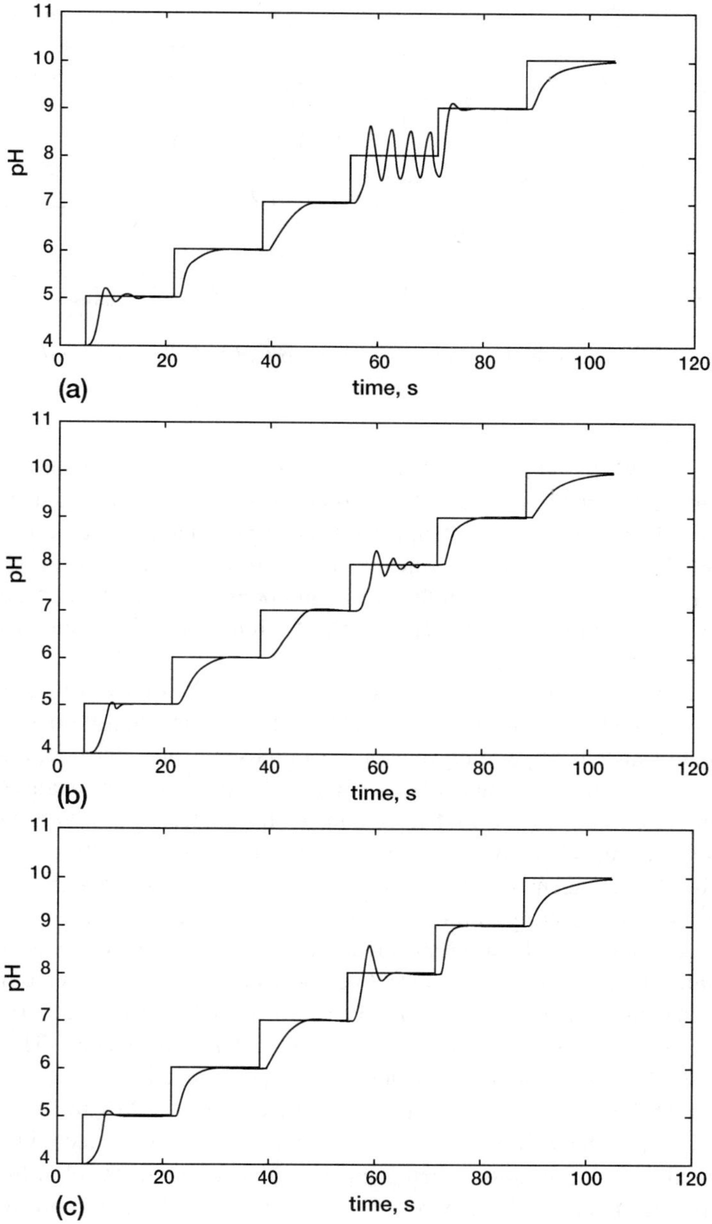

Figure 11.9 Tracking performance of (a) linear GPC, (b) LM-GPC and (c) LC-GPC controllers

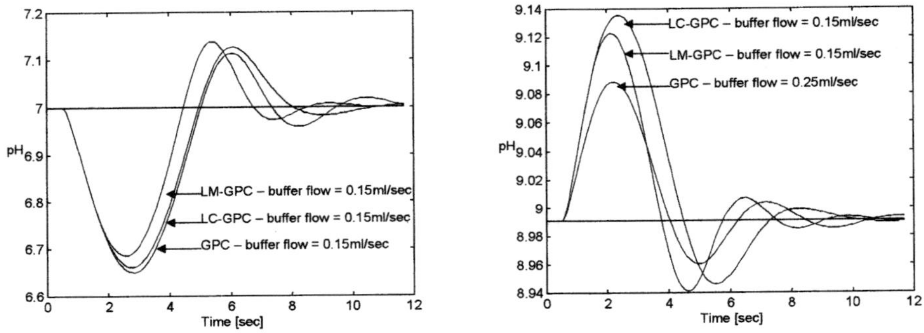

Figure 11.10 Disturbance performance of linear GPC, LM-GPC and LC-GPC controllers

Although GPC, based on a linear model, can produce excellent results compared to conventional methods such as PID, its robustness to model mismatch is an aspect which could be profitably improved. It has been argued that prediction, based solely on a single linear model, is a significant limitation of the GPC approach. It would lead to large differences between the actual and predicted output values, especially when the current output is relatively far away from the operating point at which the linear control model was generated.

In LC-GPC, a set of local controllers was combined to form a local controller network. This technique has the added attraction of improving the overall control transparency, since the operating space is decomposed using controllers rather than models. Further, the restriction of high frequency dynamics is localised to each controller rather than to the global system. Unfortunately, the number of tuning parameters was significantly increased, leading to tuning difficulties.

In LM-GPC, the accuracy of the internal model used by the nonlinear LM-GPC controller was significantly improved by utilising a nonlinear model of the process. Here a local model network representation of the pH process was shown to significantly improve predictive control performance. However, since only one set of controller tuning parameters is involved in this nonlinear control structure, the highest frequency dynamics will always have to take precedence. Therefore, in practice, the best performance may not always be achieved.

The use of the GPC control algorithm coupled with the LMN means that linear optimisation for the control moves is possible. This avoids techniques which require nonlinear optimisation when using a nonlinear neural model of the process.

Two of the main application benefits of using a predictive controller are its ability to effectively handle multivariable processes and the facility to process constraints directly within the control law. The inclusion of constraints in the above algorithms is not as straightforward as it might seem. For LM-GPC the least squares solution to the chosen cost function may be replaced by a constrained nonlinear optimisation technique such as quadratic programming, which increases the computation required to solve for the control sequence at each sample instant. The same approach could be used for the LC-GPC, whereby each controller again employs a quadratic programming technique to solve for the control sequence. However, herein lies a problem, as there is now no way of knowing that the summation of all of the controller outputs, after being passed through the validation function, will not in fact violate a process constraint. For this reason it was decided to compare both of the above algorithms without employing constraints. The LM network has also been applied to multivariate systems and there are the obvious benefits of being able to model variable interaction. However, collecting data for the identification task is not easy.

Finally, note that, while this chapter has concentrated on GPC, work has also been done on the use of LM plant models for nonlinear dynamic matrix control for pH neutralisation [40], including reported results from a laboratory-scale process [41].

11.6 References

1 LEIGH, J.R.: 'Applied control theory', IEE Control Engineering Series 18 (Peter Peregrinus, 1982)

2 LUYBEN, W.L.: 'Process modelling, simulation, and control for chemical engineers' (McGraw-Hill Chemical Engineering Series, 1990, 2nd edn)

3 SEBORG, D.E., EDGAR, T.F., and MELLICHAMP, D.A.: 'Process dynamics and control' (Wiley Series in Chemical Engineering, 1989)

4 SEBORG, D.E.: 'A perspective on advanced strategies for process control (revisited)', in FRANK, P.M. (Ed.): 'Advances in control' (Springer, 1999), pp. 103–34

5 CUTLER, C.R,. and RAMAKER, B.L.: 'Dynamic matrix control – a computer control algorithm'. Proceedings of American Control Conference, Paper WP5-B, 1980

6 RICHALET, J., RAULT, A., TESUD, L., and PAPON, J.: 'Model predictive heuristic control: Applications to industrial processes', *Automatica*, 1978, **14**, pp. 413–28

7 CLARKE, D.W., MOHTADI, C., and TUFFS, P.S.: 'Generalized predictive control. Part 1: The basic algorithm', *Automatica*, 1987, **23** (2), pp. 137–48

8 CLARKE, D.W., MOHTADI, C., and TUFFS, P.S.: 'Generalized predictive control. Part 2: Extensions and interpretations', *Automatica*, 1987, **23** (2), pp. 149–60

9 MEHRA, R.K., ROUHANI, R., ETERNO, J., RICHALET, J., and RAULT, A.: 'Model algorithmic control: review and recent developments'. Proceedings of Engineering Foundation Conference on *Chemical process control*, 1982, **2**, pp. 287–310

10 SOETERBOEK, A.R.M., VERBRUGGEN, H.B., VAN DEN BOSCH, P.P.J., and BUTLER, H.: 'On the unification of predictive control algorithms'. Proceedings of 29th Conference on *Decision and control*, 1990, pp. 1451–6

11 QIN, S.J., and BADGEWELL, T.A.: 'An overview of industrial model predictive control technology', in KANTOR, J.C., GARCIA, C.E., and

CARNAHAN, B. (Eds), Chemical Process Control – AIChE Symposium Series, 1997, **93** (316), pp. 232–56

12 QIN, S.J., and BADGEWELL, T.A.: 'An overview of nonlinear model predictive control applications'. Proceedings of IFAC Workshop on *Nonlinear model predictive control – assessment and future directions*, 1998

13 HERNANDEZ, E., and ARKUN, Y.: 'Neural network modelling and an extended DMC algorithm to control nonlinear systems'. Proceedings of American Control Conference, 1990, **3**, pp. 2454–9

14 SU, H., and McEVOY, T.J.: 'Neural model predictive control of nonlinear chemical processes'. Proceedings of international symposium on *Intelligent control*, 1993, pp. 358–63

15 TEMENGO, K.O., SCHNELLE, P.D., and McEVOY, T.J.: 'Model predictive control of an industrial packed bed reactor using neural nets', *J. Process Control*, 1995, **5** (1), pp. 19–27

16 TURNER, P., MONTAGUE, G.A., MORRIS, A.J., AGAMMENONI, O., PRITCHARD, C., BARTON, G., and ROMAGNOLI, J.: 'Application of a model based predictive control scheme to a distillation column using neural networks'. Proceedings of American Control Conference, 1995, **3**, pp. 2312–6

17 GOMM, J.B., EVANS, J.T., and WILLIAMS, D.: 'Development and performance of a neural network predictive controller', *Control Engineering Practice*, 1997, **5** (1), pp. 49–60

18 JOHANSEN T.A., and FOSS B.A.: 'Identification of nonlinear system structure and parameters using regime decomposition', *Automatica*, 1995, **31** (2), pp. 321–6

19 NELLES O.: 'Orthonormal basis functions for nonlinear system identification with linear local model trees (LOLIMOT)'. Proceedings of 11th IFAC symposium on *System identification*, Japan, 1997, **2**, pp. 667–72

20 McGINNITY, S., and IRWIN, G.W.: 'Comparison of two approaches for multiple-model identification of a pH neutralisation process'. Proceedings of European Control Conference, Paper Id-F267 (CD Rom), 1999

21 JOHANSEN, T.A., and MURRAY-SMITH, R.: 'Multiple model approaches to modelling and control' (Taylor and Francis, 1997)

22 JOHANSEN, T.A., and FOSS, B.A.: 'Constructing NARMAX models using ARMAX models', *Int. J. Control*, 1993, **58** (5), pp. 1125–53

23 JOHANSEN, T.A.: 'Operating regime based process modelling and identification'. Technical report 94–109–W, Dr. Ing. thesis, Department of Engineering Cybernetics, Norwegian Institute of Technology, Trondheim, Norway, 1994

24 FRIEDMAN, J.H.: 'Multivariate adaptive regression splines', *Annals of Statistics*, 1991, **19**, pp. 1–141

25 JOHANSEN, T.A., and FOSS, B.A.: 'Semi-empirical modelling of non-linear dynamic systems through identification of operating regimes and local models'. Advances in neural networks for control systems (Springer Verlag, 1995) pp. 105–26

26 JOHANSEN, T.A., and FOSS, B.A.: 'Operating regime based process modelling and identification', *Computers and Chemical Engineering*, 1997, **21** (2), pp. 159–76

27 NELLES, O., HECKER, O., and ISERMANN, R.: 'Automatic model selection in local linear model trees (LOLIMOT) for nonlinear system identification of a

transport delay process'. Proceedings of 11th IFAC symposium on *System identification*, Japan, 1997, **2**, pp. 727–32
28 JOHANSEN, T.A., FOSS, B.A., and SORENSEN, A.V.: 'Non-linear predictive control using local models – applied to a batch fermentation process', *Control Eng. Practice*, 1995, **3** (3), pp. 389–96
29 IRWIN, G.W., BROWN, M., and LIGHTBODY, G.: 'Nonlinear identification of a turbogenerator by local model networks'. UKACC international conference on *Control* '98, September 1998, pp. 1462–7
30 McLOONE, S.F., BROWN, M.D., IRWIN, G.W., and LIGHTBODY, G.: 'A hybrid linear/nonlinear training algorithm for feedforward neural networks', *IEEE Trans. on Neural Networks*, 1998, **9** (4), pp. 669–84
31 RUGH, W.J., and SHAMMA, J.S.: 'Research on gain scheduling', *Automatica*, 2000, **36**, pp. 1401–25
32 LEITH, D.J., and LEITHEAD, W.E.: 'Comments on "Gain scheduling dynamic linear controllers for a nonlinear plant"', *Automatica*, 1998, **34** (8), 1041–3
33 McLOONE, S., and IRWIN, G.W.: 'Process dynamical modelling using continuous time local model networks'. Proceedings of 5th IFAC workshop on *Algorithms and architectures for real-time control*, Cancun, Mexico, 1998, pp. 179–84
34 LEITH, D.J., and LEITHEAD, W.E.: 'Gain-scheduled controller design: an analytic framework directly incorporating non-equilibrium plant dynamics', *Int. J. Control*, 1998, **70** (2), pp. 249–69
35 NAHAS, E.P., HENSON, M.A., and SEBORG, D.E.: 'Nonlinear internal model control strategy for neural network models', *Computers Chem. Eng.*, 1992, **16** (12), pp. 1039–57
36 GEMAN, S., BEINENSTOCK, E., and DOURSAT, R.: 'Neural networks and the bias/variance dilemma', *Neural Computation*, 1992, **4**, pp. 1–58
37 SKRJANC, I., KAVSEK-BIASIZZO, K., and MATKO, D.: 'Fuzzy predictive control based on relational matrix models', *Computers Chem. Eng.*, 1996, **20**, pp. S931–S936
38 FISCHER, M., SCHMIDT, M., and KAVSEK-BIASIZZO, K.: 'Nonlinear predictive control based on the extraction of step response models from Takagi-Sugeno fuzzy systems'. Proceedings of ACC, Albuquerque, 1997
39 SHORTEN, R., and MURRAY-SMITH, R.: 'Side-effects of normalising basis functions', in 'Multiple model approaches to modelling and control', (Taylor and Francis, 1997), pp. 211–28
40 TOWNSEND, S., LIGHTBODY, G., BROWN, M., and IRWIN, G.W.: 'Nonlinear dynamic matrix control using local models', *Trans. Inst. Meas. Control*, 1998, **20** (1), pp. 47–56
41 DRAEGER, A., ENGELL, S., and ROSMANN, V.: 'A comparison of different classes of neural networks for predictive control of a neutralisation plant'. Preprints IFAC Conference on *Control systems design*, Slovak Republic, pp. 465–70, 2000

Chapter 12

Neural network control of a gasoline engine with rapid sampling

Barry Lennox and Gary Montague

Abstract

Artificial neural networks provide a modelling approach that enables accurate nonlinear model formulation from system data. Unfortunately, the neural network models do not fit easily within a control framework, due to computational overheads, particularly when there is a requirement for a rapid sampling frequency. This chapter demonstrates how a neural network model may be built and incorporated within a model predictive control framework and, with some approximations, implemented on a system requiring frequent sampling. Application to a realistic simulation of a gasoline engine serves to demonstrate the potential of the approach.

12.1 Introduction

Model based predictive control (MPC) has been extremely successful in its application to industrial processes. A recent survey indicated that there were over 2000 commercial applications of the technology [1]. The vast majority of these applications were implemented using linear dynamic models within the control structure. Unfortunately, there exist many control problems that contain nonlinearities of such magnitude that the application of linear MPC becomes inappropriate. Such problems include pH regulation and the control of batch processes. In an attempt to solve such control problems, many investigators have focused on the application of nonlinear control systems and in particular nonlinear MPC.

Nonlinear MPC is typically achieved by simply replacing the linear model that is used in more conventional MPC with a nonlinear model. The type of nonlinear model that has been employed in industrial applications varies and includes artificial neural networks (ANN), nonlinear state space models and polynomial functions [2]. Unfortunately the integration of a nonlinear model within the MPC algorithm produces a cost function that cannot be optimised explicitly, as is the case with unconstrained linear MPC. Instead the cost function must be optimised at each sampling interval using a nonlinear optimisation procedure, such as quasi-Newton, genetic algorithms or successive quadratic programming. With the recent improvement in modern computing, optimisation problems can be solved in a matter of seconds. Although such times are trivial compared with the sampling rates in many industrial processes, there exist many systems where such times make the application of such controllers unsuitable. This contribution analyses the application of nonlinear MPC to one such application, the control of air-fuel ratio (AFR) in a gasoline engine.

The dynamics of gasoline engines are such that sampling times are typically of the order of 10 ms, rendering standard nonlinear MPC techniques unsuitable. In this chapter it is shown that with some approximations it is possible to implement an MPC algorithm, based upon a nonlinear ANN model, that is suitable for application to the control of AFR in gasoline engines.

12.2 Artificial neural networks

The term artificial neural network encompasses many different model structures or architectures [3], the choice of which is dependent on the purpose of developing the neural network. Despite the differences that exist between network architectures, all artificial neural networks do possess some common features. They are generally composed of many neuron-like processing elements (nodes) which are heavily interconnected to form a network. The most commonly used processing element is one in which input signals to the neuron are weighted and then summed together with a bias term. The neuron output is then obtained by passing the summed, weighted inputs through a nonlinear activation function. The nonlinear function is usually sigmoidal; in this work the hyperbolic tangent function was used.

For the mapping of nonlinear systems, a layered architecture referred to as the feedforward network or multilayered perceptron is typically used. This type of network comprises an input layer where input information is presented to the network, one or more 'hidden' layers where neuron processing takes place and an output layer from which the network outputs are obtained. The topology of such layered architectures is usually described according to the number of nodes in each layer. For example, a network with two inputs, one hidden layer with four nodes and one output is referred to as a 2–4–1 network.

The basic feedforward network using sigmoidal activation functions has been shown to be able to approximate any nonlinear function to an arbitrary degree of accuracy by using either one or two hidden layers [4, 5]. Several techniques have been proposed to aid the determination of the number of hidden layers and nodes [6] but, due to their complexity, an experimental approach incorporating experience and heuristic knowledge is often preferred. In this study a manual cross-validation procedure was adopted for the determination of ANN topologies.

The feedforward network described above is useful for many applications but a number of modifications have been proposed to improve model performance. In particular in this application it is essential that system dynamics are captured by the model. One approach is to adopt the philosophy of the ARMAX and NARMAX models and use a time series of system inputs/outputs. Although this can produce reasonable dynamic models [7] there are more effective means of dynamic modelling with ANNs. An elegant approach to modelling dynamics can be performed at a localised level by incorporating first order, lowpass filters into the neuron processing elements in what is referred to as a filter based network. The values of the filter time constants are not known *a priori* and therefore they must be determined by the training algorithm, in addition to the network weights, when the neural network is trained.

The problem of dealing with the long and variable delay between system inputs and outputs is generally best tackled externally to the network. Studies have addressed the ability of a network to cope internally with modelling a system which experiences time delay variation [8] and, although small variations can be coped with (essentially by the network implementing a form of Padé approximation to a delay), large variations can cause problems. Thus in this application an external time delay model was used to time shift data before it entered the network. Since the time delay arises from flow process, such a delay model is relatively straightforward to construct. Any small errors in delay will, of course, be accommodated for by the ANN.

Once a network architecture has been specified, the network must be trained. A set of process data is selected as a training data set and the network weights, and in the case of the filter based network, the filter time constants, are determined so that the neural network represents an accurate model of the system being studied.

The issue of training is essentially a nonlinear optimisation problem, the aim being to minimise a cost function based on the sum of the squared prediction error of the ANN model. Traditionally the most popular algorithm for training basic feedforward networks has been the backpropagation algorithm, a steepest descent routine [9]. Second order optimisation techniques are known to be superior in terms of speed but have high computational requirements, restricting their use with large networks. For this study, however, very large networks were not encountered and the second order, Levenberg-Marquardt search direction method [10] was successfully employed.

Due to the complexity of the neural network architecture and the large number of adjustable parameters, it is possible to minimise prediction errors greatly and hence fit training data sets with extreme accuracy. This can occur to such a degree that the neural network model will begin to fit secondary system characteristics such as noise and measurement errors. A network trained to such accuracy will be too specific to the training data set and will generalise poorly when applied to other plant data. In order to prevent this, some form of model validation is often employed during the training procedure. In this study, at regular intervals in the optimisation routine, the neural network was validated using another data set. This involved testing the prediction accuracy of the trained model on a testing data set and continuing training until the testing data set prediction error began to increase. At this point the network was beginning to overfit the training data and therefore training was terminated.

12.3 ANN engine model development

The controller studies described in this chapter are based upon the control of a realistic simulation of a car engine which includes dynamic elements such as the filling of the intake manifold, the wall wetting of the fuel in the intake ports, the transient response of the sensors and the effect of cold start and warm-up operation including temperature effects on volumetric efficiency and wall wetting dynamics. A simplified block diagram of the engine is shown in Figure 12.1.

Over recent years a number of researchers have investigated specific aspects of the air-fuel control problem. A common area of study has been the use of adaptive systems to cope both with system nonlinearities and long term drift [11–14]. Typically, the models used in the adaptive systems are linear in form and rely on parameter updating to track the changing dynamics. To place the severity of the nonlinearities in context, it is possible for the gains and the time constants to vary by two orders of magnitude during typical operation.

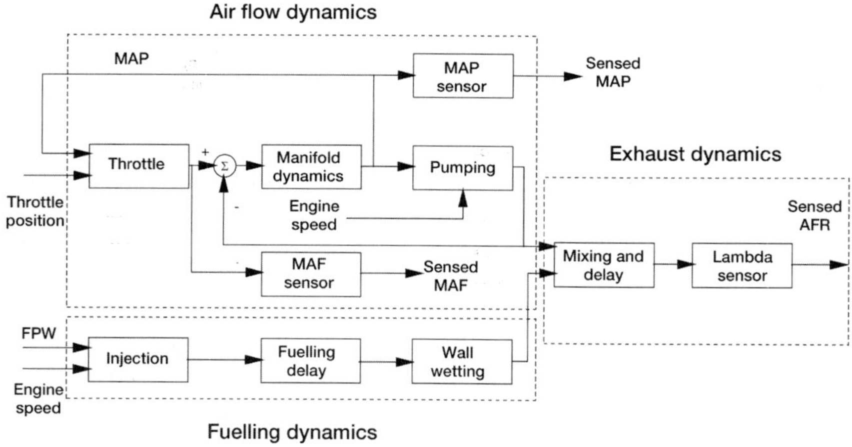

Figure 12.1 Engine dynamic model. FPW = fuel pulse width; AFR = air/fuel ratio; MAF = mass air flow; MAP = manifold air pressure

The engine model used here is a very well established model, extensive details of which can be found in Beaumont *et al.* [11]. Experimental results have also shown that a model-based controller developed using this model was capable of controlling a Peugeot 605 engine [15], thus providing experimental validation of the model.

Before controller studies could be considered, ANN models of the system had to be generated from system data. Several sets of data were obtained from the simulation by applying multilevel pseudorandom sequences (PRS) to the throttle, resulting in step changes in throttle position at random times. Fuel pulse width (FPW) was controlled from measured mass air flowrate (MAF) with an additional PRS applied to introduce an uncorrelated element between MAF and FPW. Each set of data comprised of 100 s of engine running sampled at 100 Hz. The data was sampled and a realistic level of quantisation was simulated. Three of these sets of data were then used to train, test and validate the ANN model of the process.

The quality of fit over a representative section of the validating data set is shown in Figure 12.2. It can be seen that an accurate model has been obtained that has captured both system dynamics and steady states. This figure also shows the model fit produced by a linear finite impulse response model (FIR) [16], which proved to be more accurate than both ARX and ARMAX models. Clearly the accuracy of the ANN far exceeds that of the linear FIR model. The inputs to each of these models

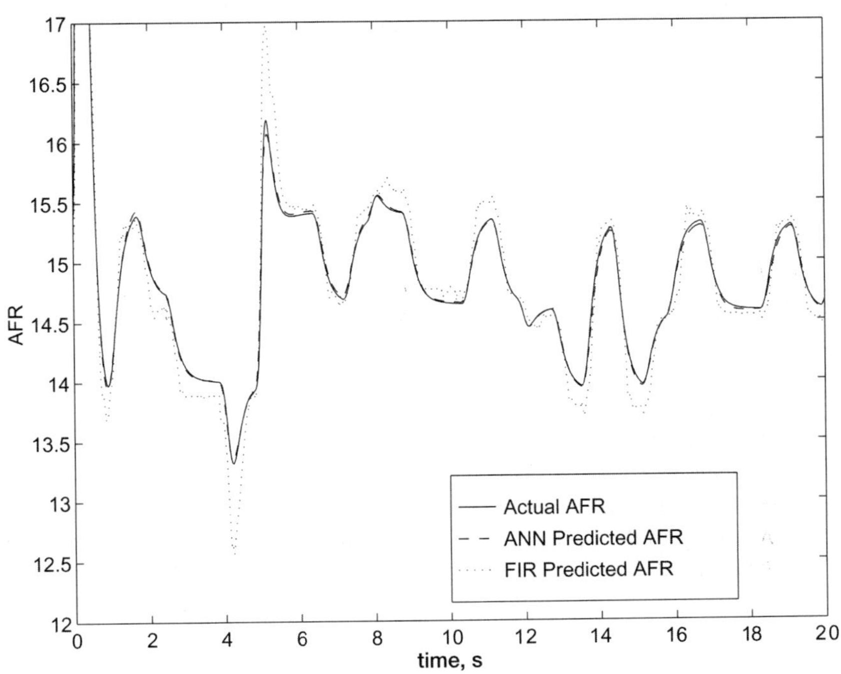

Figure 12.2 Network model fit to validating data

were the throttle position, MAF, manifold air pressure, FPW and engine speed. The optimal neural network configuration was found to contain six nodes in a single hidden layer.

12.4 Neural network based control

The work described in this chapter exploits the concepts of predictive control to overcome the time delays present in the gasoline engine and the modelling capabilities of the neural network to overcome the system nonlinearities. The basic philosophy of predictive control is to use a model of the system to predict process outputs into the future, necessitating the postulation of future control moves. By placing appropriate limits on the horizons of predictions and values that these predictions can take, it is possible to design a robust controller that is capable of handling process constraints. A number of references exist to the development of predictive control, for example Reference 17. If a predictive controller for a single output system is considered then typically a linear model is used to predict the output trajectory between horizons N_1, the time delay of the system, and N_2, often taken to be the settling time of the system. These predictions are used to determine suitable control moves up to a horizon N_u in the future. Since the model of the process is taken as linear, the control actions required to satisfy a performance criteria (J) of the form

$$J = \sum_{n=N_1}^{N_2} (w(t+n) - \hat{y}(t+n))^2 + \sum_{n=1}^{N_u} \lambda(\Delta u(t+n-1))^2 \qquad (1)$$

can be directly calculated, resulting in a control action expression:

$$\tilde{\mathbf{u}} = (\mathbf{G}^T\mathbf{G} + \lambda\mathbf{I})^{-1}\mathbf{G}^T(\mathbf{w} - \mathbf{f}) \qquad (2)$$

where $\tilde{\mathbf{u}}$ is a vector containing changes in the control action, Δu, up to the horizon N_u. Only the first of these control moves is implemented and the calculations are repeated at the next time step. $\mathbf{w}$ is a vector containing future values of the setpoint, λ is a control action weighting chosen to prevent excessive control moves, $\hat{y}$ is the model output and $\mathbf{I}$ is the identity matrix. $\mathbf{G}$ is a matrix which is based upon the step response of the process and is used to calculate the effect that future control actions will have on the process output. $\mathbf{f}$ is a vector containing future predictions of the process output up to the horizon, N_2, in the future based upon known signals. The model output can therefore be expressed as follows:

$$\hat{\mathbf{y}} = \mathbf{G}\tilde{\mathbf{u}} + \mathbf{f} \qquad (3)$$

A controller of this form has been applied to the air-fuel ratio control problem and is now discussed.

The use of a nonlinear model in the predictive control scheme introduces a problem in that the cost function (1) can only be minimised through an iterative approach, as a direct solution of the form of (2) is not possible. Although convergence to a solution is not guaranteed, the use of a nonlinear optimisation algorithm, such as quasi-Newton [10] to calculate future control moves which minimise the cost function is relatively straightforward, and constraints in input variable changes are easily dealt with by using a constrained optimisation algorithm [18]. If constraints are present for the output then augmenting the cost function with a high cost for constraint violation is possible. This procedure is discussed in Montague *et al.* [19]. The major drawback in using this iterative controller formulation is that the optimisation algorithm takes time to converge to a solution. While for many processes the calculation time is trivial (say of the order of a second), if the controller must sample every 0.01s then such a solution strategy becomes inappropriate.

To enable the rapid minimisation of the cost of (1), the controller used in these studies uses the neural network in two ways: one to calculate the elements of the vector **f** and secondly to generate a linear approximation of the process step response around the operating conditions that the model predicts will occur at the prediction horizon N_1 in the future. This linear approximation will form the step response matrix **G** and will be calculated along with **f** at each sampling instant.

The parameters of **G** are determined by simulating the effect of a small step change around the predicted operating point at $t + N_1 - 1$ and using a simple least squares fit to parameterise the linear model. Thus by making a linear approximation to the nonlinear step response at some time in the future a direct solution for the control action is possible. Since the operating conditions will change at each sampling instant then so too will the elements of **G**. Therefore **G** must be calculated along with **f** at each sampling instant. A similar technique to this has been proposed by Lightbody and Irwin [20] who solved a neural network based internal model control structure by linearising the nonlinear model. Gattu and Zafiriou [21] also applied a similar technique to control a semi-batch polymerisation reactor using quadratic DMC with a nonlinear state space model.

The expectation would be that minimising the cost of (1) using a nonlinear optimiser would provide better control system performance than linearising the system at each sampling instant. By approximating the response between $t + N_1 - 1$ and $t + N_2$ with a linear model around the current operating point the control system performance would be expected to be degraded slightly. Where the time delay is large (as in the case of air-fuel ratio control) there will be a significant difference between this error and that which would arise by making the linear approximation around the operating point at time $t - 1$. All these controllers would of course also require a nonlinear model. If a fixed linear approximation of the process over the whole operating range were to be assumed then the control performance could be expected to be poor compared to nonlinear model based controllers, if the process is significantly nonlinear. These expectations were investigated by application of the controllers to the engine air-fuel ratio control system.

12.4.1 Application of the ANN model based controller to the gasoline engine

To determine whether there is a significant degradation in controller performance between a linear approximation of the step response of the ANN model and solving the controller cost function using a nonlinear optimiser (termed here iterative solution), both types of controller were installed onto the engine simulation. The values for N_2, N_u and λ were manipulated to ensure the best possible control performance, measured in terms of the ability of the controller to regulate the AFR during the throttle disturbance without using excessive control action. The size of this disturbance was a change in throttle position from 30° to 40° (the throttle is fully closed at approximately 0° and fully open at approximately 90°) over a period of 0.1 s. Figure 12.3 compares the performance of the two controllers with N_2 equal to 10 and N_u equal to 1.

This figure demonstrates that there is little difference between the two controllers, and the approximation technique used in this study would appear to be suitable for regulation of AFR in the gasoline engine.

As a final comparison, the performance of the controller using the linear approximation to the ANN model was compared with a fixed linear model based predictive controller. For both cases the values of N_2 and N_u were 10 and 1, respectively. Figure 12.4 shows the performance of the two controllers for the same step change in throttle position investigated earlier. This figure shows that there is

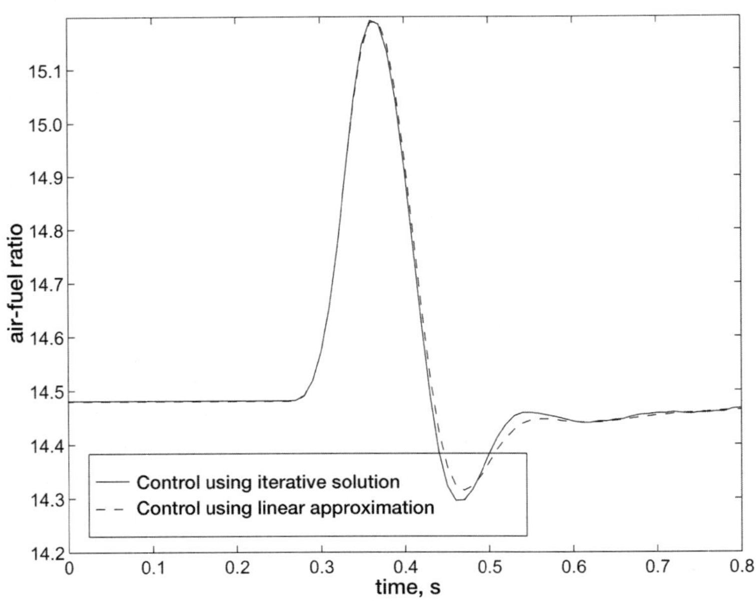

Figure 12.3 Comparison of the control achieved through an iterative solution with that achieved by linearising the ANN model locally

enormous benefit in using the ANN rather than linear model based predictive controller. The reason for this improvement is because the linear model is not capable of predicting the nonlinearities which are present in this highly complex process. Increasing the value of N_2 was found to have no improvement on the accuracy of the control obtained with the fixed linear model based predictive controller.

12.5 Conclusions

This contribution has investigated the use of an artificial neural network model for the control of a system that required frequent sampling. It has been shown that with some application approximations it is possible to gain the benefits provided by a nonlinear system model and a sophisticated control strategy without sacrificing performance significantly. The technique for the implementation of an AFR control strategy using an artificial neural network model has been described which allows for very fast computation of controller actions. It is therefore a feasible approach for application.

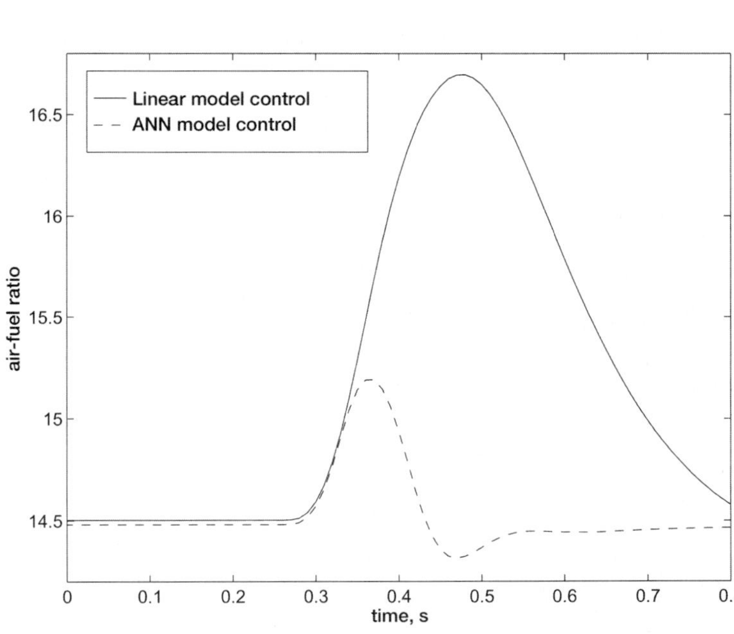

Figure 12.4 Comparison of linear and nonlinear model based predictive control

12.6 References

1 QIN, S.J., and BADGWELL, T.A.: 'An overview of industrial model predictive control technology'. Fifth international conference on *Chemical process control*, AIChE and CACHE, pp. 232–56

2 QIN, S.J., and BADGWELL, T.A.: 'An overview of nonlinear model predictive control applications' in ALLGÖWER, F. and ZHENG, A. (Eds): 'Nonlinear model predictive control' (Birkhauser, 2000)

3 LIPPMAN, R.P.: 'An introduction to computing with neural nets', *IEEE ASSP Magazine*, April 1987, pp. 4–22

4 CYBENKO., G.: 'Approximations by superpositions of a sigmoidal function', *Math. Control Signals Syst.*, 1989, **2**, pp. 303–14

5 HORNIK, K., STINCHOMBE, M., and WHITE, H.: 'Multilayer feed-forward networks are universal approximators', *Neural Networks*, 1989, **2**, pp. 359–66

6 PAGE, G.F., GOMM, J.B., and WILLIAMS, D. (Eds), 'Application of neural networks to modelling and control' (Chapman & Hall, 1993)

7 BHAT, N., and MCAVOY, T.J.: 'Use of neural nets for dynamic modelling and control of chemical process systems', *Comput. Chem. Eng.*, 1990, **14** (4/5), pp. 573–83

8 TURNER, P., LENNOX, B., MONTAGUE, G.A., and MORRIS, A.J.: 'Modelling complex systems using artificial neural networks'. World Congress on Neural networks, San Diego, 1996

9 RUMMELHART, D.E., and MCCLELLAND, J.L.: 'Parallel distributed processing: explorations in the microstructure of cognition' (MIT Press, Cambridge, MA, 1986, vol. 1)

10 SCALES, L.E.: 'Introduction to nonlinear optimisation' (Macmillan Publishers Ltd, 1985)

11 BEAUMONT, A.J., NOBLE, A.D., and SCARISBRICK, A.: 'Adaptive transient air-fuel ratio control to minimise gasoline engine emissions'. FISITA Congress, London, 1992

12 INGAKI, H., OHATA, A., and INOUC, T.: 'An adaptive fuel injection control with internal model in automotive engines'. IECON, Monterey, 1993

13 AULT, B.A., JONES, V.K., POWELL, J.D., and FRANKLIN, G.F.: 'Adaptive air fuel ratio control of a spark-ignition engine', 1994, SAE 940373

14 TURIN, R.C., and GEERING, H.P.: 'Model based adaptive fuel control in an SI engine'. SAE 940374, 1994

15 BEAUMONT, A.J., and NOBLE, A.D.: 'Air fuel ratio control technology for ultra low emissions vehicle'. Paper 93EN032, 26th ISATA, *The motor vehicle and the environment – demands of the nineties and beyond*, 1993

16 LJUNG, L.: 'System identification – theory for the user' (Prentice Hall, Englewood Cliffs, NJ, 1987)

17 CLARKE, D.W., MOHTADI, C., and TUFFS, P.S.: 'Generalised predictive control. Part 1: The basic algorithm and Part 2: Extensions and interpretations', *Automatica*, 1987, **23** (2), pp. 137–60

18 DE OLIVEIRA, N.M.C., and BIEGLER, L.T.: 'Constraint handling and stability properties of model-predictive control', *AIChE J.*, 1994, **40** (7), pp. 1138–55

19 MONTAGUE, G.A., THAM, M.T., WILLIS, M.J., and MORRIS, A.J.: 'Predictive control of distillation columns using dynamic neural networks'. Dycord+ '92, University of Maryland, College Park, 1992
20 LIGHTBODY, G., and IRWIN, G.: 'Novel neural network internal model control structure'. Proceedings of the American Control Conference, 1995, **1**, pp. 350–4
21 GATTU, G., and ZAFIRIOU, E.: 'Nonlinear quadratic dynamic matrix control with state estimation', *Ind. Eng. Chem. Res.*, **31** (4), pp. 1096–104

Index